CliffsQuickReview™ Calculus

Anton/Bivens/Davis Version

By Bernard V. Zandy, MA and
Jonathan J. White, MS

About the Authors

Bernard V. Zandy, MA, Professor of Mathematics at Fullerton College in California has been teaching secondary and college level mathematics for 34 years. A co-author of the Cliffs PSAT and SAT Preparation Guides, Mr. Zandy has been a lecturer and consultant for Bobrow Test Preparation Services, conducting workshops at California State University and Colleges since 1977.

Jonathan J. White has a BA in mathematics from Coe College and an MS in mathematics from the University of Iowa. He is currently pursuing a PhD in Mathematics Pedagogy and Curriculum Research at the University of Oklahoma.

Publisher's Acknowledgments

Editorial

Project Editor: Brian Kramer
Acquisitions Editor: Sherry Gomoll
Technical Editor: Dale Johnson

Composition

Indexer: TECHBOOKS Production Services
Proofreader: Joel K. Draper
Wiley Publishing Composition Services

CliffsQuickReview™ Calculus Anton/Bivens/Davis Version
Published by
Learneo, Inc.
www.learneo.com
www.cliffsnotes.com

Library of Congress Control Number available from the Library of Congress

ISBN: 978-1-957671-20-8

Printed in the United States of America

10 9 8 7 6 5 4 3 2 1

1O/QS/QW/QT/IN

Published by Learneo, Inc.

Table of Contents

CORRELATION GUIDE: *CLIFFSQUICKREVIEW CALCULUS* WITH *CALCULUS 7E* BY ANTON, BIVENS, DAVIS

CliffsQuickReview Calculus		***Calculus 7E***
Chapter	***Topic Title***	
1	Interval Notation	Appendix A
1	Absolute Value	Appendix B
1	Functions	Sections 1.1, 1.2
1	Linear Equations	Appendix C
1	Trigonometric Functions	Appendix E
2	Intuitive Definition	Section 2.1
2	Evaluating Limits	Section 2.2
2	One-Sided Limits	Sections 2.1, 2.2
2	Infinite Limits	Sections 2.1, 2.2
2	Limits at Infinity	Sections 2.1, 2.3
2	Limits Involving Trigonometric Functions	Section 2.6
2	Continuity	Section 2.5
3	Definition	Section 3.2
3	Differentiation Rules	Section 3.3
3	Trigonometric Function Differentiation	Section 3.4
3	Chain Rule	Section 3.5
3	Implicit Differentiation	Section 3.6
3	Higher Order Derivatives	Section 3.3
3	Differentiation of Inverse Trigonometric Functions	Section 7.6
3	Differentiation of Exponential and Logarithmic Functions	Section 7.3

(continued)

CliffsQuickReview Calculus		***Calculus 7E***
Chapter	***Topic Title***	
4	Tangent and Normal Lines	Section 3.2
4	Critical Points	Section 4.2
4	Extreme Value Theorem	Section 4.5
4	Mean Value Theorem	Section 4.8
4	Increasing/Decreasing Functions	Section 4.1
4	First Derivative Test for Local Extrema	Section 4.2
4	Second Derivative Test for Local Extrema	Section 4.2
4	Concavity and Points of Inflection	Section 4.1
4	Maximum/Minimum Problems	Section 4.6
4	Distance, Velocity, Acceleration	Section 4.4
4	Related Rates of Change	Section 3.7
4	Differentials	Section 3.8
5	Antiderivatives/Indefinite Integrals	Section 5.2
5	Integration Techniques	Section 5.2
5	Substitution and Change of Variables	Section 5.3
5	Integration by Parts	Section 8.2
5	Trigonometric Integrals	Section 8.3
5	Trigonometric Substitutions	Section 8.4
5	Distance, Velocity, and Acceleration	Section 5.7
5	Definition of Definite Integrals	Section 5.5
5	Properties of Definite Integrals	Sections 5.5, 5.6
5	The Fundamental Theorem of Calculus	Section 5.6
5	Definite Integral Evaluation	Section 5.8

Chapter	Topic Title	
6	Area	Section 6.1
6	Volumes of Solids with Known Cross Sections	Section 6.2
6	Disk Method	Section 6.2
6	Washer Method	Section 6.2
6	Cylindrical Shell Method	Section 6.3
6	Arc Length	Section 6.4

CORRELATION GUIDE: *CLIFFSQUICKREVIEW CALCULUS* WITH *CALCULUS: EARLY TRANSCENDENTALS 7E* BY ANTON, BIVENS, DAVIS

CliffsQuickReview Calculus		*ET Calculus 7E*
Chapter	**Topic Title**	
1	Interval Notation	Appendix A
1	Absolute Value	Appendix B
1	Functions	Sections 1.1, 1.2
1	Linear Equations	Appendix C
1	Trigonometric Functions	Appendix E
2	Intuitive Definition	Section 2.1
2	Evaluating Limits	Section 2.2
2	One-Sided Limits	Sections 2.1, 2.2
2	Infinite Limits	Sections 2.1, 2.2
2	Limits at Infinity	Sections 2.1, 2.3

(continued)

CliffsQuickReview Calculus		***ET Calculus 7E***
Chapter	***Topic Title***	
2	Limits Involving Trigonometric Functions	Section 2.6
2	Continuity	Section 2.5
3	Definition	Section 3.2
3	Differentiation Rules	Section 3.3
3	Trigonometric Function Differentiation	Section 3.4
3	Chain Rule	Section 3.5
3	Implicit Differentiation	Section 3.6
3	Higher Order Derivatives	Section 3.3
3	Differentiation of Inverse Trigonometric Functions	Section 4.4
3	Differentiation of Exponential and Logarithmic Functions	Section 4.3
4	Tangent and Normal Lines	Section 3.2
4	Critical Points	Section 5.2
4	Extreme Value Theorem	Section 5.5
4	Mean Value Theorem	Section 5.8
4	Increasing/Decreasing Functions	Section 5.1
4	First Derivative Test for Local Extrema	Section 5.2
4	Second Derivative Test for Local Extrema	Section 5.2
4	Concavity and Points of Inflection	Section 5.1
4	Maximum/Minimum Problems	Section 5.6
4	Distance, Velocity, Acceleration	Section 5.4
4	Related Rates of Change	Section 3.7
4	Differentials	Section 3.8

Chapter	Topic Title	
5	Antiderivatives/Indefinite Integrals	Section 6.2
5	Integration Techniques	Section 6.2
5	Substitution and Change of Variables	Section 6.3
5	Integration by Parts	Section 8.2
5	Trigonometric Integrals	Section 8.3
5	Trigonometric Substitutions	Section 8.4
5	Distance, Velocity, and Acceleration	Section 6.7
5	Definition of Definite Integrals	Section 6.5
5	Properties of Definite Integrals	Sections 6.5, 6.6
5	The Fundamental Theorem of Calculus	Section 6.6
5	Definite Integral Evaluation	Section 6.8
6	Area	Section 7.1
6	Volumes of Solids with Known Cross Sections	Section 7.2
6	Disk Method	Section 7.2
6	Washer Method	Section 7.2
6	Cylindrical Shell Method	Section 7.3
6	Arc Length	Section 7.4

CQR Pocket Guide
Calculus

Geometry and Algebra

Area, Circumference, and Volume:

Rectangle $A = bh$

Triangle $A = \frac{1}{2}bh$

Circle $A = \pi r^2 \quad C = 2\pi r$

Sphere $V = \frac{4}{3}\pi r^3 \quad A = 4\pi r^2$

Cylinder $V = \pi r^2 h \quad A = 2\pi rh + 2\pi r^2$

Cone $V = \frac{1}{3}\pi r^2 h$

Quadratic Formula:

$$x = \frac{-b \pm \sqrt{b^2 - 4ac}}{2a}$$

Factoring:

$x^2 - y^2 = (x - y)(x + y)$

$x^3 - y^3 = (x - y)(x^2 + xy + y^2)$

$x^3 + y^3 = (x + y)(x^2 - xy + y^2)$

Lines:

$m = \frac{y_2 - y_1}{x_2 - x_1}$

$y = mx + b$

$y - y_1 = m(x - x_1)$

$Ax + By + C = 0$

Exponents:

$x^n \cdot x^m = x^{n+m}$

$\frac{x^n}{x^m} = x^{n-m}$

$(x^n)^m = x^{nm}$

Logarithms:

$\log_a(xy) = \log_a x + \log_a y$

$\log_a(x/y) = \log_a x - \log_a y$

$\log_a(x^n) = n \log_a x$

Trigonometry

Trig values for selected angles:

rad	*deg*	*sin x*	*cos x*	*tan x*
0	0°	0	1	0
$\pi/6$	30°	½	$\sqrt{3}/2$	$\sqrt{3}/3$
$\pi/4$	45°	$\sqrt{2}/2$	$\sqrt{2}/2$	1
$\pi/3$	60°	$\sqrt{3}/2$	½	$\sqrt{3}$
π	90°	1	0	–

Key Trig Identities:

$\sin^2 \Theta + \cos^2 \Theta = 1$

$\tan^2 \Theta + 1 = \sec^2 \Theta$

$1 + \cot^2 \Theta = \csc^2 \Theta$

$\sin(x + y) = \sin x \cos y + \cos x \sin y$

$\sin(x - y) = \sin x \cos y - \cos x \sin y$

$\cos(x + y) = \cos x \cos y - \sin x \sin y$

$\cos(x - y) = \cos x \cos y + \sin x \sin y$

$\sin 2\Theta = 2\sin \Theta \cos \Theta$

$\cos 2\Theta = \cos^2 \Theta - \sin^2 \Theta$

$= 2\cos^2 \Theta - 1$

$= 1 - 2\sin^2 \Theta$

Law of Sines:

$$\frac{\sin A}{a} = \frac{\sin B}{b} = \frac{\sin C}{c}$$

Law of Cosines:

$c^2 = a^2 + b^2 - 2ab \cos C$

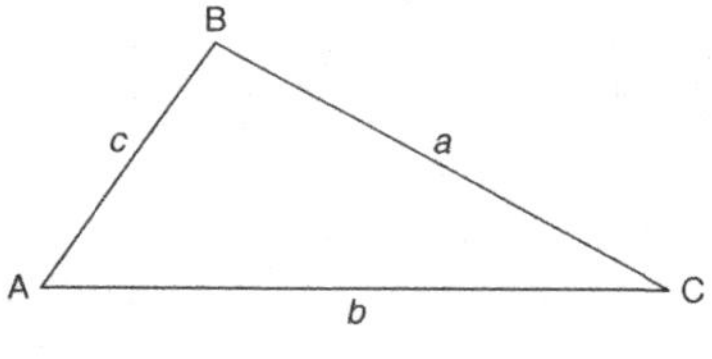

CQR Pocket Guide
Calculus

Derivatives

$$\frac{d}{dx}\left(f\cdot g\right)=f\cdot g'+g\cdot f'$$

$$\frac{d}{dx}\left(\frac{f}{g}\right)=\frac{g\cdot f'-f\cdot g'}{g^2}$$

$$\frac{d}{dx}\left[f\left(g\right)\right]=f'\left(g\right)\cdot g'$$

$$\frac{d}{dx}(x^n)=nx^{n-1}$$

$$\frac{d}{dx}(\sin x)=\cos x$$

$$\frac{d}{dx}(\cos x)=-\sin x$$

$$\frac{d}{dx}(\tan x)=\sec^2 x$$

$$\frac{d}{dx}(\cot x)=-\csc^2 x$$

$$\frac{d}{dx}(\sec x)=\sec x\tan x$$

$$\frac{d}{dx}(\csc x)=-\csc x\cot x$$

$$\frac{d}{dx}(e^x)=e^x$$

$$\frac{d}{dx}(a^x)=(\ln a)\,a^x$$

$$\frac{d}{dx}\left(\ln x\right)=1/x$$

$$\frac{d}{dx}\left(\log_a x\right)=\frac{1}{(\ln a)\cdot x}$$

$$\frac{d}{dx}\left(\sin^{-1}x\right)=\frac{1}{\sqrt{1-x^2}}$$

$$\frac{d}{dx}\left(\cos^{-1}x\right)=\frac{-1}{\sqrt{1-x^2}}$$

$$\frac{d}{dx}\left(\tan^{-1}x\right)=\frac{1}{1+x^2}$$

$$\frac{d}{dx}\left(\csc^{-1}x\right)=\frac{-1}{|x|\sqrt{x^2-1}}$$

$$\frac{d}{dx}\left(\sec^{-1}x\right)=\frac{1}{|x|\sqrt{x^2-1}}$$

$$\frac{d}{dx}\left(\cot^{-1}x\right)=\frac{-1}{1+x^2}$$

Antiderivatives

$$\int u\,dv=uv-\int v\,du$$

$$\int k\,dx=kx+C$$

$$\int x^n\,dx=\frac{x^{n+1}}{n+1}+C,\ n\neq -1$$

$$\int \sin x\,dx=-\cos x+C$$

$$\int \cos x\,dx=\sin x+C$$

$$\int \sec^2 x\,dx=\tan x+C$$

$$\int \csc^2 x\,dx=-\cot x+C$$

$$\int \sec x\tan x\,dx=\sec x+C$$

$$\int \csc x\cot x\,dx=-\csc x+C$$

$$\int e^x\,dx=e^x+C$$

$$\int a^x\,dx=\frac{a^x}{\ln a}+C,\ a>0,\ a\neq 1$$

$$\int \frac{dx}{x}=\ln|x|+C$$

$$\int \tan x\,dx=-\ln|\cos x|+C$$

$$\int \cot x\,dx=\ln|\sin x|+C$$

$$\int \sec x\,dx=\ln|\sec x+\tan x|+C$$

$$\int \csc x\,dx=-\ln|\csc x+\cot x|+C$$

$$\int \frac{dx}{\sqrt{a^2-x^2}}=\sin^{-1}\frac{x}{a}+C$$

$$\int \frac{dx}{a^2+x^2}=\frac{1}{a}\tan^{-1}\frac{x}{a}+C$$

$$\int \frac{dx}{x\sqrt{x^2-a^2}}=\frac{1}{a}\sec^{-1}\left|\frac{x}{a}\right|+C$$

For more information about Wiley,
call 1-800-762-2974.

Introduction

Calculus is the mathematics of change. Any situation that involves quantities that change over time can be understood with the tools of calculus. **Differential calculus** deals with rates of change or slopes, and is explored in Chapters 3 and 4 of this book. **Integral calculus** handles total changes or areas, and is addressed in Chapters 5 and 6. Although it is not always immediately obvious, this mathematical notion of change is essential to many areas of knowledge, particularly disciplines like physics, chemistry, biology, and economics.

The prerequisites for learning calculus include much of high school algebra and trigonometry, as well as some essentials of geometry. If the formulas on the front side of the Pocket Guide (the cardstock page right inside the front cover) and topics covered in Chapter 1 are familiar to you, then you probably have sufficient background to begin learning calculus. If some of those are unfamiliar, or just rusty for you, then *CliffsQuickReview Geometry, CliffsQuickReview Algebra,* or *CliffsQuickReview Trigonometry* may be valuable starting points for you.

Why You Need This Book

Can you answer yes to any of these questions?

- Do you need to review the fundamentals of calculus fast?
- Do you need a course supplement to calculus?
- Do you need a concise, comprehensive reference for calculus?

If so, then *CliffsQuickReview Calculus* is for you!

How to Use This Book

You can use this book in any way that fits your personal style for study and review—you decide what works best with your needs. You can either read the book from cover to cover or just look for the information you want and put it back on the shelf for later. Here are just a few ways you can use this book:

- Read the book as a stand-alone textbook to learn all the major concepts of calculus.

- Use the Pocket Guide to find often-used formulas, from calculus and other relevant formulas from algebra, geometry and trigonometry.
- Refer to a single topic in this book for a concise and understandable explanation of an important idea.
- Get a glimpse of what you'll gain from a chapter by reading through the "Chapter Check-In" at the beginning of each chapter.
- Use the Chapter Checkout at the end of each chapter to gauge your grasp of the important information you need to know.
- Test your knowledge more completely in the CQR Review and look for additional sources of information in the CQR Resource Center.
- Review the most important concepts of an area of calculus for an exam.
- Brush up on key points as preparation for more advanced mathematics.

Using Calculus 7e by Anton/Bivens/Davis (ABD) with CQR

- ABD to CQR. If you are reading the ABD text, use the Correlation Guides in the front of the CQR (right before page 1) to quickly find the corresponding topic in your CliffsQuickReview.
- CQR to ABD. If you are using your CQR, you can easily find additional explanation or examples in Anton/Bivens/Davis—some section heads within this CliffsQuickReview are followed by an icon indicating on what page you can find more help in Anton/Bivens/Davis. In cases when there are two icons, the icon containing "ET" indicates that the section number differs in the Early Transcendentals version of the book for that reference.

Being a valuable reference source also means it's easy to find the information you need. Here are a few ways you can search for topics in this book:

- Look for areas of interest in the book's Table of Contents, or use the index to find specific topics.
- Use the glossary to find key terms fast. This book defines new terms and concepts where they first appear in the chapter. If a word is boldfaced, you can find a more complete definition in the book's glossary.
- Flip through the book looking for subject areas at the top of each page.

Chapter 1
REVIEW TOPICS

Chapter Check-In

- ❑ Reviewing functions
- ❑ Using equations of lines
- ❑ Reviewing trigonometric functions

Certain topics in algebra, geometry, analytical geometry, and trigonometry are essential in preparing to study calculus. Some of them are briefly reviewed in the following sections.

Interval Notation

The set of real numbers (*R*) is the one that you will be most generally concerned with as you study calculus. This set is defined as the union of the set of rational numbers with the set of irrational numbers. Interval notation provides a convenient abbreviated notation for expressing intervals of real numbers without using inequality symbols or set-builder notation.

The following lists some common intervals of real numbers and their equivalent expressions, using set-builder notation:

$$(a, b) = \{x \in R: a < x < b\}$$

$$[a, b] = \{x \in R: a \leq x \leq b\}$$

$$[a, b) = \{x \in R: a \leq x < b\}$$

$$(a, b] = \{x \in R: a < x \leq b\}$$

$$(a, +\infty) = \{x \in R: x > a\}$$

$$[a, +\infty) = \{x \in R: x \geq a\}$$

$$(-\infty, b) = \{x \in R: x < b\}$$

$$(-\infty, b] = \{x \in R: x \leq b\}$$

$$(-\infty, +\infty) = \{x \in R\}$$

Note that an infinite end point ($\pm\infty$) is never expressed with a bracket in interval notation because neither $+\infty$ nor $-\infty$ represents a real number value.

Absolute Value

The concept of absolute value has many applications in the study of calculus. The absolute value of a number a, written $|a|$ may be defined in a variety of ways. On a real number line, the absolute value of a number is the distance, disregarding direction, that the number is from zero. This definition establishes the fact that the absolute value of a number must always be nonnegative—that is, $|a| \geq 0$.

A common algebraic definition of absolute value is often stated in three parts, as follows:

$$|a| = \begin{cases} a, a > 0 \\ 0, a = 0 \\ -a, a < 0 \end{cases}$$

Another definition that is sometimes applied to calculus problems is

$$|a| = \sqrt{a^2}$$

or the principal square root of a^2. Each of these definitions also implies that the absolute value of a number must be a nonnegative.

Functions

A **function** is defined as a set of ordered pairs (x,y), such that for each first element x, there corresponds one and only one second element y. The set of first elements is called the *domain* of the function, while the set of second elements is called the *range* of the function. The domain variable is referred to as the independent variable, and the range variable is referred to as the dependent variable. The notation $f(x)$ is often used in place of y to indicate the value of the function f for a specific replacement for x and is read "f of x" or "f at x."

Geometrically, the graph of a set or ordered pairs (x,y) represents a function if any vertical line intersects the graph in, at most, one point. If a vertical line were to intersect the graph at two or more points, the set would have one x value corresponding to two or more y values, which clearly contradicts the definition of a function. Many of the key concepts and theorems of calculus are directly related to functions.

Example 1-1: The following are some examples of equations that define functions.

(a) $y = f(x) = 3x + 1$
(b) $y = f(x) = x^2$
(c) $y = f(x) = |x| - 5$
(d) $y = f(x) = -3$
(e) $y = f(x) = \dfrac{x-3}{x^2+4}$
(f) $y = f(x) = \sqrt[3]{2x+9}$
(g) $y = f(x) = \dfrac{6}{x}$
(h) $y = \tan x$
(i) $y = \cos 2x$

Example 1-2: The following are some equations that do not define functions; each has an example to illustrate why it does not define a function.

(a) $x = y^2$; If $x = 4$, then $y = 2$ or $y = -2$
(b) $x = |y + 3|$; If $x = 2$, then $y = -5$ or $y = -1$
(c) $x = -5$; If $x = -5$, then y can be any real number.
(d) $x^2 + y^2 = 25$; If $x = 0$, then $y = 5$ or $y = -5$.
(e) $y = \pm\sqrt{x+4}$; If $x = 5$, then $y = +3$ or $y = -3$.
(f) $x^2 - y^2 = 9$; If $x = -5$, then $y = 4$ or $y = -4$.

Linear Equations App C

A **linear equation** is any equation that can be expressed in the form $Ax + By + C = 0$, where A and B are not both zero. Although a linear equation may not be expressed in this form initially, it can be manipulated algebraically to this form. The slope of a line indicates whether the line slants up or down to the right or is horizontal. The slope is usually denoted by the letter m and is defined in a number of ways:

$$\begin{aligned} m &= \frac{\text{rise}}{\text{sun}} \\ &= \frac{\text{vertical change}}{\text{horizontal change}} \\ &= \frac{y \text{ value change}}{x \text{ value change}} \\ &= \frac{\Delta y}{\Delta x} \\ &= \frac{y_2 - y_1}{x_2 - x_1} \\ &= \frac{y_1 - y_2}{x_1 - x_2} \end{aligned}$$

Note that for a vertical line, the x value would remain constant, and the horizontal change would be zero; hence, a vertical line is said to have no slope or its slope is said to be **nonexistent** or **undefined.** All nonvertical lines have a numerical slope with a positive slope indicating a line slanting up to the right, a negative slope indicating a line slanting down to the right, and a slope of zero indicating a horizontal line.

Example 1-3: Find the slope of the line passing through (–5, 4) and (–1, –3).

$$\begin{aligned} m &= \frac{y_2 - y_1}{x_2 - x_1} \\ &= \frac{(-3) - (4)}{(-1) - (-5)} \\ &= -\frac{7}{4} \end{aligned}$$

The line, then, has a slope of –7/4.

Some forms of expressing linear equations are given special names that identify how the equations are written. Note that even though these forms appear to be different from one another, they can be algebraically manipulated to show they are equivalent.

Any nonvertical lines are parallel if they have the same slopes, and conversely lines with equal slopes are parallel. If the slopes of two lines L_1 and L_2 are m_1 and m_2, respectively, then L_1 is parallel to L_2 if and only if $m_1 = m_2$.

Two nonvertical, nonhorizontal lines are perpendicular if the product of their slopes is –1, and conversely, if the product of their slopes is –1, the lines are perpendicular. If the slopes of two lines L_1 and L_2 are m_1 and m_2, respectively, then L_1 is perpendicular to L_2 if and only if $m_1 \cdot m_2 = -1$.

You should note that any two vertical lines are parallel and a vertical line and a horizontal line are always perpendicular.

The **general** or **standard form** of a linear equation is $Ax + By + C = 0$, where A and B are not both zero. If $B = 0$, the equation takes the form x = constant and represents a vertical line. If $A = 0$, the equation takes the form y = constant and represents a horizontal line.

Example 1-4: The following are some examples of linear equations expressed in general form:

(a) $2x + 5y - 10 = 0$

(b) $x - 4y = 0$

(c) $x + 3 = 0$

(d) $y - 6 = 0$

The **point-slope form** of a linear equation is $y - y_1 = m(x - x_1)$ when the line passes through the point (x_1, y_1) and has a slope m.

Example 1-5: Find an equation of the line through the point (3,4) with slope –2/3.

$$y - y_1 = m(x - x_1)$$

$$y - 4 = -\frac{2}{3}(x - 3)$$

$$y - 4 = -\frac{2}{3}x + 2$$

$$y = -\frac{2}{3}x + 6$$

$$3y = -2x + 18$$

$$2x + 3y - 18 = 0 \quad \text{(general form)}$$

The **slope-intercept form** of a linear equation is $y = mx + b$ when the line has y-intercept $(0, b)$ and slope m.

Example 1-6: Find an equation of the line that has a slope 4/3 and crosses y-axis at –5.

$$y = mx + b$$

$$y = \frac{4}{3}x + (-5)$$

$$3y = 4x - 15$$

$$4x - 3y - 15 = 0 \quad \text{(general form)}$$

The **double intercept form** of a linear equation is $x/a + y/b = 1$ when the line has x-intercept $(a,0)$ and y-intercept $(0,b)$.

Example 1-7: Find the double intercept and slope intercept form of an equation of the line that crosses the x-axis at –2 and the y-axis at 3.

$$\frac{x}{a} + \frac{y}{b} = 1$$

$$\frac{x}{-2} + \frac{y}{3} = 1 \text{ (double intercept form)}$$

$$y = \frac{3}{2}y + 3 \text{ (slope intercept form)}$$

Trigonometric Functions

In trigonometry, angle measure is expressed in one of two units: degrees or radians. The relationship between these measures may be expressed as follows: $180^\circ = \pi$ radians.

To change degrees to radians, the equivalent relationship $1^\circ = \pi/180$ radians is used, and the given number of degrees is multiplied by $\pi/180$ to convert to radian measure. Similarly, the equation 1 radian $= 180^\circ/\pi$ is used to change radians to degrees by multiplying the given radian measure by $180/\pi$ to obtain the degree measure.

The six basic trigonometric functions may be defined using a circle with equation $x^2 + y^2 = r^2$ and the angle θ in standard position with its vertex at the center of the circle and its initial side along the positive portion of the x-axis (see Figure 1-1).

The trigonometric functions sine, cosine, tangent, cotangent, secant, and cosecant are defined as follows:

Figure 1-1 Defining the trigonometric functions.

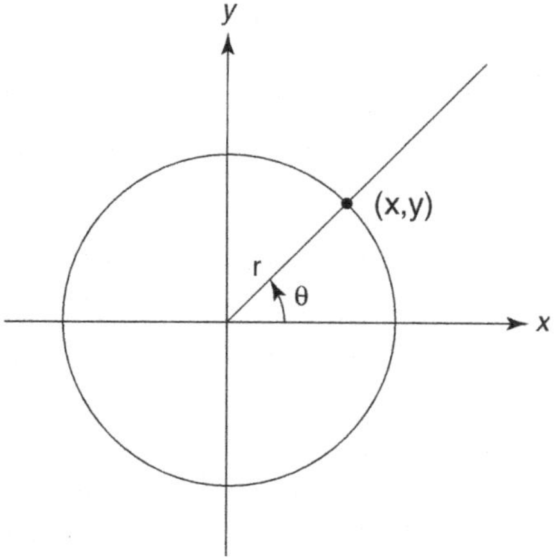

$$\sin\theta = \frac{y}{r} = \frac{y}{\sqrt{x^2+y^2}}$$

$$\cos\theta = \frac{x}{r} = \frac{x}{\sqrt{x^2+y^2}}$$

$$\tan\theta = \frac{\sin\theta}{\cos\theta} = \frac{y}{x}$$

$$\cot\theta = \frac{\cos\theta}{\sin\theta} = \frac{x}{y}$$

$$\sec\theta = \frac{1}{\cos\theta} = \frac{r}{x} = \frac{\sqrt{x^2+y^2}}{x}$$

$$\csc\theta = \frac{1}{\sin\theta} = \frac{r}{y} = \frac{\sqrt{x^2+y^2}}{y}$$

It is essential that you be familiar with the values of these functions at multiples of 30°, 45°, 60°, 90°, and 180° (or in radians, $\pi/6$, $\pi/4$, $\pi/3$, $\pi/2$, and π (See Table 1-1.) You should also be familiar with the graphs of the six trigonometric functions. Some of the more common trigonometric identities that are used in the study of calculus are as follows:

$\sin^2\theta + \cos^2\theta = 1$

$\tan^2\theta + 1 = \sec^2\theta$

$1 + \cot^2\theta = \csc^2\theta$

$\sin(-\theta) = -\sin\theta$

$\cos(-\theta) = \cos\theta$

$\tan(-\theta) = -\tan\theta$

$\sin(\theta + 2\pi) = \sin\theta$

$\cos(\theta + 2\pi) = \cos\theta$

$\tan(\theta + \pi) = \tan\theta$

$\sin(A + B) = \sin A\cos B + \cos A\sin B$

$\sin(A - B) = \sin A\cos B - \cos A\sin B$

$\cos(A + B) = \cos A\cos B - \sin A\sin B$

$\cos(A - B) = \cos A\cos B + \sin A\sin B$

$$\tan(A + B) = \frac{\tan A + \tan B}{1 - \tan A\tan B}$$

$$\tan(A - B) = \frac{\tan A - \tan B}{1 + \tan A\tan B}$$

$\sin 2\theta = 2\sin\theta\cos\theta$

$$\begin{aligned}\cos 2\theta &= \cos^2\theta - \sin^2\theta\\ &= 2\cos^2\theta - 1\\ &= 1 - 2\sin^2\theta\end{aligned}$$

$$\sin^2\frac{1}{2}\theta = \frac{1 - \cos\theta}{2}$$

$$\cos^2\frac{1}{2}\theta = \frac{1 + \cos\theta}{2}$$

The relationship between the angles and sides of a triangle may be expressed using the **Law of Sines** or the **Law of Cosines** (see Figure 1-2).

Figure 1-2 Relations between sides and angles of a triangle.

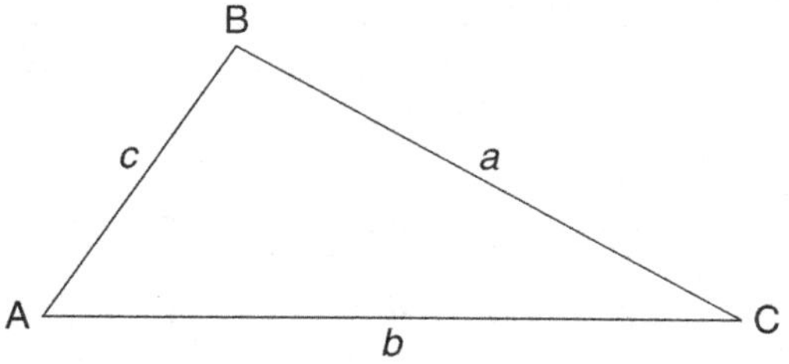

Law of Sines: $\frac{\sin A}{a} = \frac{\sin B}{b} = \frac{\sin C}{c}$

or $\frac{a}{\sin A} = \frac{b}{\sin B} = \frac{c}{\sin C}$

Law of Cosines: $a^2 = b^2 + c^2 - 2bc \cos A$

or $b^2 = a^2 + c^2 - 2ac \cos B$

or $c^2 = a^2 + b^2 - 2ab \cos C$

Table 1-1 Values of Sine, Cosine, and Tangent at Common Angles

Degree Measure of x	Radian Measure of x	sin x	cos x	tan x
0	0	0	1	0
30	$\frac{\pi}{6}$	$\frac{1}{2}$	$\frac{\sqrt{3}}{2}$	$\frac{\sqrt{3}}{3}$
45	$\frac{\pi}{4}$	$\frac{\sqrt{2}}{2}$	$\frac{\sqrt{2}}{2}$	1
60	$\frac{\pi}{3}$	$\frac{\sqrt{3}}{2}$	$\frac{1}{2}$	$\sqrt{3}$
90	$\frac{\pi}{2}$	1	0	Undefined
120	$\frac{2\pi}{3}$	$\frac{\sqrt{3}}{2}$	$-\frac{1}{2}$	$-\sqrt{3}$
135	$\frac{3\pi}{4}$	$\frac{\sqrt{2}}{2}$	$-\frac{\sqrt{2}}{2}$	-1
150	$\frac{5\pi}{6}$	$\frac{1}{2}$	$-\frac{\sqrt{3}}{2}$	$-\frac{\sqrt{3}}{3}$
180	π	0	-1	0
210	$\frac{7\pi}{6}$	$-\frac{1}{2}$	$-\frac{\sqrt{3}}{2}$	$\frac{\sqrt{3}}{3}$
225	$\frac{5\pi}{4}$	$-\frac{\sqrt{2}}{2}$	$-\frac{\sqrt{2}}{2}$	1
240	$\frac{4\pi}{3}$	$-\frac{\sqrt{3}}{2}$	$-\frac{1}{2}$	$\sqrt{3}$
270	$\frac{3\pi}{2}$	-1	0	Undefined
300	$\frac{5\pi}{3}$	$-\frac{\sqrt{3}}{2}$	$\frac{1}{2}$	$-\sqrt{3}$
315	$\frac{7\pi}{4}$	$-\frac{\sqrt{2}}{2}$	$\frac{\sqrt{2}}{2}$	-1
330	$\frac{11\pi}{6}$	$-\frac{1}{2}$	$\frac{\sqrt{3}}{2}$	$-\frac{\sqrt{3}}{3}$
360	2π	0	1	0

Chapter Checkout

Q&A

1. Which of the following equations does not define a function?
   a. $3x - 2y = 6$
   b. $y = \sin 3x$
   c. $y = x^2$
   d. $x^2 + y^2 = 16$
   e. $y = \sqrt{3 + x}$
2. Find an equation in general form of the line with slope $-2/5$ and y-intercept $(0,3)$.
3. Find an equation in general form of the line passing through the origin and perpendicular to the line $5x - 3y = 6$.
4. Find an equation in general form of the line passing through the points $(3,-2)$ and $(-1,0)$.
5. If θ is an angle between $\pi/2$ and π, and $\sin \theta = 3/5$, what is $\cos \theta$?

Answers: 1. d **2.** $2x + 5y - 15 = 0$ **3.** $3x + 5y = 0$ **4.** $x + 2y + 1 = 0$ **5.** $-4/5$

Chapter 2

LIMITS

Chapter Check-In

- ❑ Understanding what limits are
- ❑ Computing limits
- ❑ Determining when a function is continuous

The concept of the limit of a function is essential to the study of calculus. It is used in defining some of the most important concepts in calculus—continuity, the derivative of a function, and the definite integral of a function.

Intuitive Definition

The **limit** of a function $f(x)$ describes the behavior of the function close to a particular x value. It does not necessarily give the value of the function at x. You write $\lim_{x \to a} f(x) = L$, which means that as x "approaches" a, the function $f(x)$ "approaches" the real number L (see Figure 2-1).

Figure 2-1 The limit of f(x) as x approaches a.

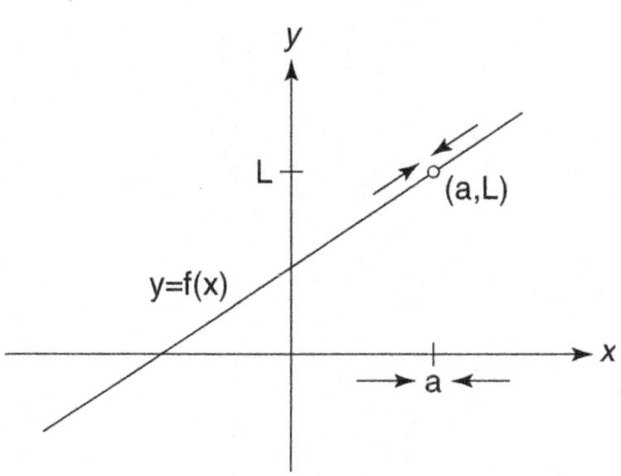

In other words, as the independent variable x gets closer and closer to a, the function value $f(x)$ gets closer to L. Note that this does not imply that $f(a) = L$; in fact, the function may not even be defined at a (Figure 2-2) or may equal some value different than L at a (Figure 2-3).

If the function does not approach a real number L as x approaches a, the limit does not exist; therefore, you write $\lim_{x \to a} f(x)$ DNE (Does Not Exist). Many different situations could occur in determining that the limit of a function does not exist as x approaches some value.

Figure 2-2 *f(a)* does not exist, but $\lim_{x \to a} f(x)$ does.

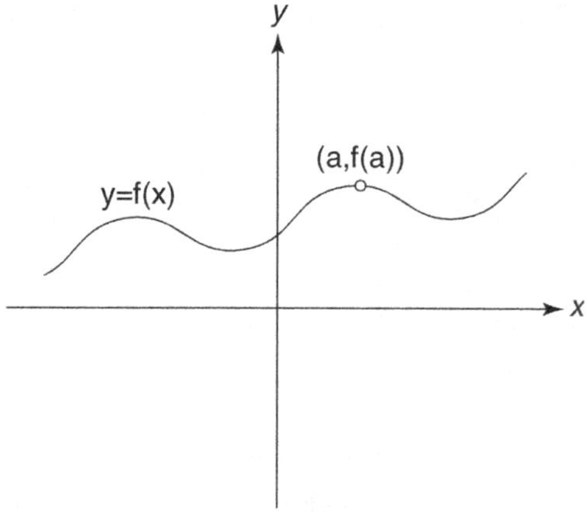

Figure 2-3 *f(a)* and $\lim_{x \to a} f(x)$ are not equal.

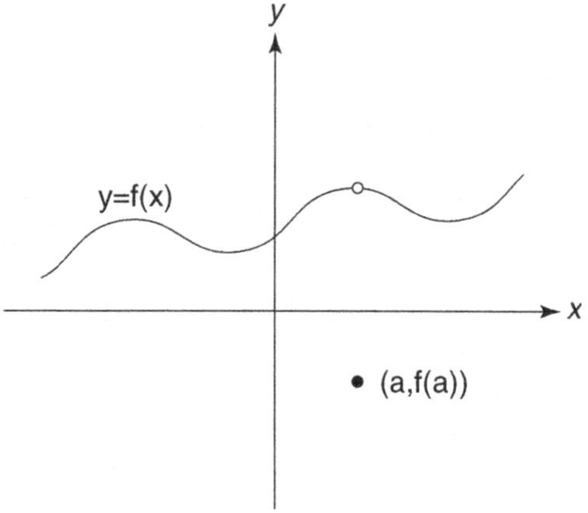

Evaluating Limits

Limits of functions are evaluated using many different techniques such as recognizing a pattern, simple substitution, or using algebraic simplifications. Some of these techniques are illustrated in the following examples.

Example 2-1: Find the limit of the sequence: $\frac{1}{2}, \frac{2}{3}, \frac{3}{4}, \frac{4}{5}, \frac{5}{6}, \frac{6}{7}, \ldots$

Because the value of each fraction gets slightly larger for each term, while the numerator is always one less than the denominator, the fraction values will get closer and closer to 1; hence, the limit of the sequence is 1.

Example 2-2: Evaluate $\lim_{x \to 2} (3x - 1)$.

As x approaches 2, $3x$ approaches 6, and $3x - 1$ approaches 5; hence, $\lim_{x \to 2} (3x - 1) = 5$.

Example 2-3: Evaluate $\lim_{x \to -3} \frac{x^2 - 9}{x + 3}$.

Substituting –3 for x yields 0/0, which is meaningless. Factoring first and simplifying, you find that

$$\begin{aligned} \lim_{x \to -3} \frac{x^2 - 9}{x + 3} &= \lim_{x \to -3} \frac{(x + 3)(x - 3)}{x + 3} \\ &= \lim_{x \to -3} (x - 3) \\ &= -6 \end{aligned}$$

The graph of $(x^2 - 9)/(x + 3)$ would be the same as the graph of the linear function $y = x - 3$ with the single point (–3,–6) removed from the graph (see Figure 2-4).

Figure 2-4 The graph of $y = (x^2 - 9)/(x + 3)$.

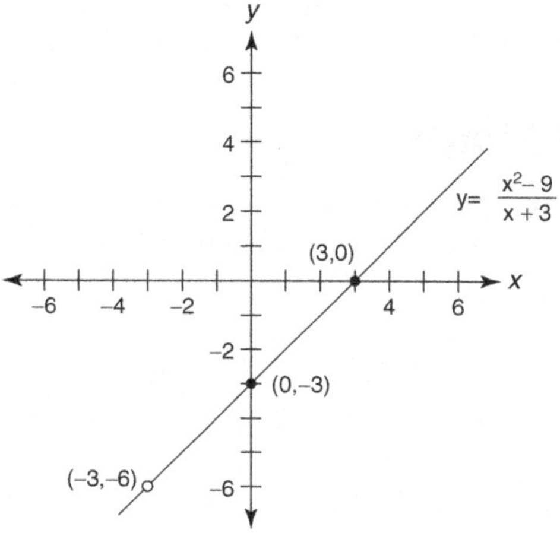

Example 2-4: Evaluate $\lim_{x \to 3} \frac{\frac{x}{x+2} - \frac{3}{5}}{x - 3}$.

Substituting 3 for x yields 0/0, which is meaningless. Simplifying the compound fraction, you find that

$$\lim_{x \to 3} \frac{\frac{x}{x+2} - \frac{3}{5}}{x - 3} = \lim_{x \to 3} \frac{\frac{x}{x+2} - \frac{3}{5}}{x - 3} \cdot \frac{5(x+2)}{5(x+2)}$$

$$= \lim_{x \to 3} \frac{5x - 3(x+2)}{5(x+2)(x-3)}$$

$$= \lim_{x \to 3} \frac{2x - 6}{5(x+2)(x-3)}$$

$$= \lim_{x \to 3} \frac{2(x-3)}{5(x+2)(x-3)}$$

$$= \lim_{x \to 3} \frac{2}{5(x+2)}$$

$$= \frac{2}{25}$$

Example 2-5: Evaluate $\lim_{x \to 0} \frac{x}{x+5}$.

Substituting 0 for x yields $0/5 = 0$; hence, $\lim_{x \to 0} x/(x+5) = 0$.

Example 2-6: Evaluate $\lim_{x \to 0} \frac{x+5}{x}$.

Substituting 0 for x yields 5/0, which is meaningless; here, $\lim_{x \to 0} (x+5)/x$ DNE. (Remember, infinity is not a real number.)

One-sided Limits

For some functions, it is appropriate to look at their behavior from one side only. If x approaches a from the right only, you write

$$\lim_{x \to a^+} f(x)$$

or if x approaches a from the left only, you write

$$\lim_{x \to a^-} f(x)$$

It follows, then, that $\lim_{x \to a} f(x) = L$ if and only if $\lim_{x \to a^+} f(x) = \lim_{x \to a^-} f(x) = L$.

Example 2-7: Evaluate $\lim_{x \to 0^+} \sqrt{x}$.

Because x is approaching 0 from the right, it is always positive; $\sqrt{x}$ is getting closer and closer to zero, so $\lim_{x \to 0^+} \sqrt{x} = 0$. Although substituting 0 for x would yield the same answer, the next example illustrates why this technique is not always appropriate.

Example 2-8: Evaluate $\lim_{x \to 0^-} \sqrt{x}$.

Because x is approaching 0 from the left, it is always negative, and $\sqrt{x}$ does not exist. In this situation, $\lim_{x \to 0^-} \sqrt{x}$ DNE. Also, note that $\lim_{x \to 0} \sqrt{x}$ DNE because $\lim_{x \to 0^+} \sqrt{x} = 0 \neq \lim_{x \to 0^-} \sqrt{x}$.

Example 2-9: Evaluate

$$\text{(a) } \lim_{x \to 2^-} \frac{|x-2|}{x-2}$$

$$\text{(b) } \lim_{x \to 2^+} \frac{|x-2|}{x-2}$$

$$\text{(c) } \lim_{x \to 2} \frac{|x-2|}{x-2}$$

(a) As x approaches 2 from the left, $x - 2$ is negative, and $|x - 2| = -(x - 2)$; hence,

$$\lim_{x \to 2^-} \frac{|x-2|}{x-2} = \frac{-(x-2)}{x-2} = -1$$

(b) As x approaches 2 from the right, $x - 2$ is positive, and $|x - 2| = x - 2$; hence;

$$\lim_{x \to 2^-} \frac{|x-2|}{x-2} = \frac{(x-2)}{x-2} = 1$$

(c) Because $\lim_{x \to 2^-} \frac{|x-2|}{x-2} \neq \lim_{x \to 2^+}$, $\lim_{x \to 2} \frac{|x-2|}{x-2}$ DNE

Infinite Limits

Some functions "take off" in the positive or negative direction (increase or decrease without bound) near certain values for the independent variable. When this occurs, the function is said to have an *infinite limit;* hence, you write $\lim_{x \to a} f(x) = +\infty$ or $\lim_{x \to a} f(x) = -\infty$. Note also that the function has a vertical asymptote at $x = a$ if either of the above limits hold true.

In general, a fractional function will have an infinite limit if the limit of the denominator is zero and the limit of the numerator is not zero. The sign of the infinite limit is determined by the sign of the quotient of the numerator and the denominator at values close to the number that the independent variable is approaching.

Example 2-10: Evaluate $\lim_{x \to 0} \frac{1}{x^2}$.

As x approaches 0, the numerator is always positive and the denominator approaches 0 and is always positive; hence, the function increases without bound and $\lim_{x \to 0} 1/x^2 = +\infty$. The function has a vertical asymptote at $x = 0$ (see Figure 2-5).

Figure 2-5 The graph of $y = 1/x^2$.

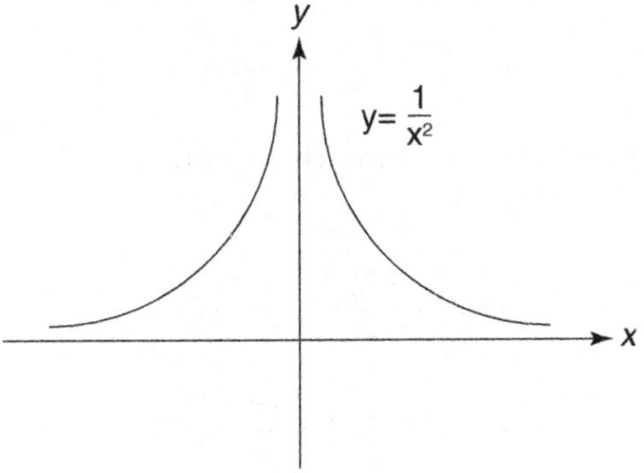

Example 2-11: Evaluate $\lim_{x \to 2^-} \frac{x+3}{x-2}$.

As x approaches 2 from the left, the numerator approaches 5, and the denominator approaches 0 through negative values; hence, the function decreases without bound and $\lim_{x \to 2^-} (x+3)/(x-2) = -\infty$. The function has a vertical asymptote at $x = 2$.

Example 2-12: Evaluate $\lim_{x \to 0^+} \left(\frac{1}{x^2} - \frac{1}{x^3} \right)$.

Rewriting $1/x^2 - 1/x^3$ as an equivalent fractional expression $(x-1)/x^3$, the numerator approaches -1, and the denominator approaches 0 through positive values as x approaches 0 from the right; hence, the function decreases without bound and $\lim_{x \to 0} \left(1/x^2 - 1/x^3 \right) = -\infty$. The function has a vertical asymptote at $x = 0$.

A word of caution: Do not evaluate the limits individually and subtract because $\pm\infty$ are not real numbers. Using this example,

$$\lim_{x \to 0^+} \left(\frac{1}{x^2} - \frac{1}{x^2} \right) \neq \lim_{x \to 0^+} \frac{1}{x^2} - \lim_{x \to 0^+} \frac{1}{x^3} = (+\infty) - (+\infty)$$

which is meaningless.

Limits at Infinity

Limits at infinity are used to describe the behavior of functions as the independent variable increases or decreases without bound. If a function approaches a numerical value L in either of these situations, write

$$\lim_{x \to +\infty} f(x) = L \text{ or } \lim_{x \to -\infty} f(x) = L$$

and $f(x)$ is said to have a horizontal asymptote at $y = L$. A function may have different horizontal asymptotes in each direction, have a horizontal asymptote in one direction only, or have no horizontal asymptotes.

Evaluate 2-13: Evaluate $\lim_{x \to +\infty} \frac{2x^2 + 3}{x^2 - 5x - 1}$.

Factor the largest power of x in the numerator from each term and the largest power of x in the denominator from each term.

You find that

$$\lim_{x \to +\infty} \frac{2x^2 + 3}{x^2 - 5x - 1} = \lim_{x \to +\infty} \frac{x^2\left(2 + \frac{3}{x^2}\right)}{x^2\left(1 - \frac{5}{x} - \frac{1}{x^2}\right)}$$

$$= \lim_{x \to +\infty} \frac{2 + \frac{3}{x^2}}{1 - \frac{5}{x} - \frac{1}{x^2}}$$

$$= \frac{2 + 0}{1 - 0 - 0}$$

$$\lim_{x \to +\infty} \frac{2x^2 + 3}{x^2 - 5x - 1} = 2$$

The function has a horizontal asymptote at $y = 2$.

Example 2-14: Evaluate $\lim_{x \to +\infty} \frac{x^3 - 2}{5x^4 - 3x^3 + 2x}$.

Factor x^3 from each term in the numerator and x^4 from each term in the denominator, which yields

$$\lim_{x \to -\infty} \frac{x^3 - 2}{5x^4 - 3x^3 + 2x} = \lim_{x \to -\infty} \frac{x^3\left(1 - \frac{2}{x^3}\right)}{x^4\left(5 - \frac{3}{x} + \frac{2}{x^3}\right)}$$

$$= \lim_{x \to -\infty} \left(\frac{1}{x}\right)\left(\frac{1 - \frac{2}{x^3}}{5 - \frac{3}{x} + \frac{2}{x^3}}\right)$$

$$= (0)\left(\frac{1-0}{5-0+0}\right)$$

$$= 0$$

The function has a horizontal asymptote at $y = 0$.

Example 2-15: Evaluate $\lim_{x \to +\infty} \frac{9x^2}{x+2}$.

Factor x^2 from each term in the numerator and x from each term in the denominator, which yields

$$\lim_{x \to +\infty} \frac{9x^2}{x+2} = \lim_{x \to +\infty} \frac{x^2(9)}{x\left(1 + \frac{2}{x}\right)}$$

$$= \lim_{x \to +\infty} x\left(\frac{9}{1 + \frac{2}{x}}\right)$$

$$= \left[\lim_{x \to +\infty} (x)\right]\left[\frac{9}{1+0}\right]$$

$$= \left[\lim_{x \to +\infty} (x)\right][9]$$

$$\lim_{x \to +\infty} \frac{9x^2}{x+2} = +\infty$$

Because this limit does not approach a real number value, the function has no horizontal asymptote as x increases without bound.

Example 2-16: Evaluate $\lim_{x \to -\infty} \left(x^3 - x^2 - 3x\right)$.

Factor x^3 from each term of the expression, which yields

$$\lim_{x \to -\infty} (x^3 - x^2 - 3x) = \lim_{x \to -\infty} (x^3)\left(1 - \frac{1}{x} - \frac{3}{x^2}\right)$$

$$= \lim_{x \to -\infty} (x^3) \cdot \lim_{x \to -\infty} \left(1 - \frac{1}{x} - \frac{3}{x^2}\right)$$

$$= \lim_{x \to -\infty} (x^3) \cdot [1 - 0 - 0]$$

$$= \left[\lim_{x \to -\infty} (x^3)\right] \cdot [1]$$

$$\lim_{x \to -\infty} (x^3 - x^2 - 3x) = -\infty$$

As in the previous example, this function has no horizontal asymptote as x decreases without bound.

Limits Involving Trigonometric Functions

The trigonometric functions sine and cosine have four important limit properties:

$$\lim_{x \to c} \sin x = \sin c$$

$$\lim_{x \to c} \cos x = \cos c$$

$$\lim_{x \to 0} \frac{\sin x}{x} = 1$$

$$\lim_{x \to 0} \frac{1 - \cos x}{x} = 0$$

You can use these properties to evaluate many limit problems involving the six basic trigonometric functions.

Example 2-17: Evaluate $\lim_{x \to 0} \frac{\cos x}{\sin x - 3}$.

Substituting 0 for x, you find that $\cos x$ approaches 1 and $\sin x - 3$ approaches –3; hence,

$$\lim_{x \to 0} \frac{\cos x}{\sin x - 3} = -\frac{1}{3}$$

Example 2-18: Evaluate $\lim_{x \to 0^+} \cot x$.

Because $\cot x = \cos x/\sin x$, you find $\lim_{x \to 0^+} \cos x/\sin x$. The numerator approaches 1 and the denominator approaches 0 through positive values because we are approaching 0 in the first quadrant; hence, the function increases without bound and $\lim_{x \to 0^+} \cot x = +\infty$, and the function has a vertical asymptote at $x = 0$.

Example 2-19: Evaluate $\lim_{x \to 0} \frac{\sin 4x}{x}$.

Multiplying the numerator and the denominator by 4 produces

$$\begin{aligned}\lim_{x \to 0} \frac{\sin 4x}{x} &= \lim_{x \to 0} \frac{4 \sin 4x}{4x} \\ &= \left(\lim_{x \to 0} 4\right) \cdot \lim_{x \to 0} \frac{\sin 4x}{4x} \\ &= 4 \cdot 1\end{aligned}$$

$$\lim_{x \to 0} \frac{\sin 4x}{x} = 4$$

Example 2-20: Evaluate $\lim_{x \to 0} \frac{\sec x - 1}{x}$.

Because $\sec x = 1/\cos x$, you find that

$$\begin{aligned}\lim_{x \to 0} \frac{\sec x - 1}{x} &= \lim_{x \to 0} \frac{\frac{1}{\cos x} - 1}{x} \\ &= \lim_{x \to 0} \frac{1 - \cos x}{x \cos x} \\ &= \lim_{x \to 0} \left(\frac{1}{\cos x}\right) \cdot \left(\frac{1 - \cos x}{x}\right) \\ &= \left[\lim_{x \to 0} \frac{1}{\cos x}\right] \cdot \left[\lim_{x \to 0} \frac{1 - \cos x}{x}\right] \\ &= 1 \cdot 0\end{aligned}$$

$$\lim_{x \to 0} \frac{\sec x - 1}{x} = 0$$

Continuity

A function $f(x)$ is said to be **continuous** at a point $(c, f(c))$ if each of the following conditions is satisfied:

(1) $f(c)$ is defined (c is in the domain of f),

(2) $\lim_{x \to c} f(x)$ exists, and

(3) $\lim_{x \to c} f(x) = f(c)$.

Geometrically, this means that there is no gap, split, or missing point for $f(x)$ at c and that a pencil could be moved along the graph of $f(x)$ through $(c, f(c))$ without lifting it off the graph. A function is said to be continuous at $(c, f(c))$ from the right if $\lim_{x \to c^+} f(x) = f(c)$ and continuous at $(c, f(c))$ from the left if $\lim_{x \to c^-} f(x) = f(c)$. Many of our familiar functions such as linear, quadratic and other polynomial functions, rational functions, and the trigonometric functions are continuous at each point in their domain. A special function that is often used to illustrate one-sided limits is the greatest integer function. The *greatest integer function,* $[x]$, is defined to be the largest integer less than or equal to x (see Figure 2-6).

Some values of $[x]$ for specific x values are

$$[2] = 2$$

$$[5.8] = 5$$

$$\left[-3\frac{1}{3}\right] = -4$$

$$[.46] = 0$$

Figure 2-6 The graph of the greatest integer function $y = [x]$.

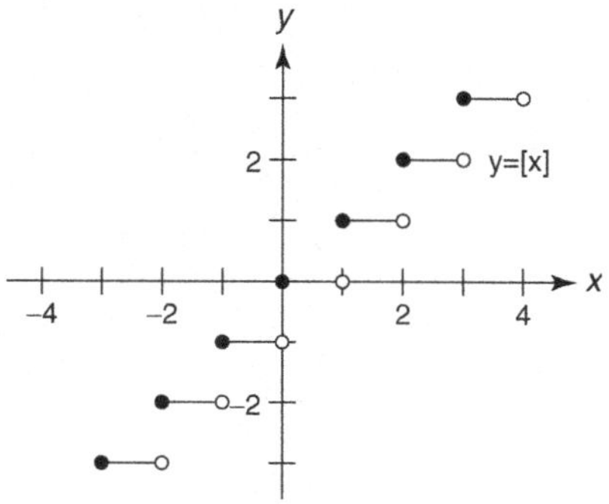

The greatest integer function is continuous at any integer n from the right only because

$$f(n) = [n] = n$$

$$\text{and } \lim_{x \to n^+} f(x) = n$$

$$\text{but } \lim_{x \to n^-} f(x) = n - 1$$

hence, $\lim_{x \to n} f(x) \neq f(n)$ and $f(x)$ is not continuous at n from the left. Note that the greatest integer function is continuous from the right and from the left at any noninteger value of x.

Example 2-21: Discuss the continuity of $f(x) = 2x + 3$ at $x = -4$.

When the definition of continuity is applied to $f(x)$ at $x = -4$, you find that

$$(1)\ f(-4) = -5$$

$$(2)\ \lim_{x \to -4} f(x) = \lim_{x \to -4} (2x + 3) = -5$$

$$(3)\ \lim_{x \to -4} f(x) = f(-4)$$

hence, f is continous at $x = -4$.

Example 2-22: Discuss the continuity of $f(x) = \frac{x^2 - 4}{x - 2}$ at $x=2$.

When the definition of continuity is applied to $f(x)$ at $x = 2$, you find that $f(2)$ is not defined; hence, f is not continuous (discontinuous) at $x = 2$.

Example 2-23: Discuss the continuity of $f(x) = \begin{cases} \frac{x^2 - 4}{x - 2}, & x \neq 2 \\ 4, & x = 2 \end{cases}$

When the definition of continuity is applied to $f(x)$ at $x = 2$, you find that

$$(1)\ f(2) = 4$$

$$\begin{aligned}(2)\ \lim_{x \to 2} f(x) &= \lim_{x \to 2} \frac{x^2 - 4}{x - 2} \\ &= \lim_{x \to 2} \frac{(x - 2)(x + 2)}{x - 2} \\ &= \lim_{x \to 2} (x + 2)\end{aligned}$$

$$\lim_{x \to 2} f(x) = 4$$

$$(3)\ \lim_{x \to 2} f(x) = f(2)$$

hence, f is continous at $x = 2$.

Example 2-24: Discuss the continuity of $f(x) = \sqrt{x}$ at $x = 0$.

When the definition of continuity is applied to $f(x)$ at $x = 0$, you find that

$$(1)\ f(0) = 0$$

$$(2)\ \lim_{x \to 0} f(x) = \lim_{x \to 0} \sqrt{x}\ \text{DNE because} \lim_{x \to 0^+} \sqrt{x} = 0,$$

$$\text{but} \lim_{x \to 0^-} \sqrt{x}\ \text{DNE}$$

$$(3)\ \lim_{x \to 0^+} f(x) = f(0)$$

hence, f is continuous at $x = 0$ from the right only.

Example 2-25: Discuss the continuity of $f(x) = \begin{cases} 5 - 2x, & x < -3 \\ x^2 + 2, & x \geq -3 \end{cases}$ at $x = -3$.

When the definition of continuity is applied to $f(x)$ at $x = -3$, you find that

(1) $f(-3) = (-3)^2 + 2 = 11$

(2) $\lim_{x \to -3^-} f(x) = \lim_{x \to -3^-} (5 - 2x) = 11$

$\lim_{x \to -3^+} f(x) = \lim_{x \to -3^+} (x^2 + 2) = 11$

hence, $\lim_{x \to 3} f(x) = 11$ because $\lim_{x \to -3^-} f(x) = \lim_{x \to -3^+} f(x)$

(3) $\lim_{x \to -3} f(x) = f(-3)$

hence, f is continuous at $x = -3$.

Many theorems in calculus require that functions be continuous on intervals of real numbers. A function $f(x)$ is said to be continuous on an open interval (a,b) if f is continuous at each point $c \in (a,b)$. A function $f(x)$ is said to be continuous on a closed interval $[a,b]$ if f is continuous at each point $c \in (a,b)$ and if f is continuous at a from the right and continuous at b from the left.

Example 2-26:

(a) $f(x) = 2x + 3$ is continuous on $(-\infty,+\infty)$ because f is continuous at every point $c \in (-\infty,+\infty)$.

(b) $f(x) = (x - 3)/(x + 4)$ is continuous on $(-\infty,-4)$ and $(-4,+\infty)$ because f is continuous at every point $c \in (-\infty,-4)$ and $c \in (-4,+\infty)$

(c) $f(x) = (x - 3)/(x + 4)$ is not continuous on $(-\infty,-4]$ or $[-4,+\infty)$ because f is not continuous on -4 from the left or from the right.

(d) $f(x) = \sqrt{x}$ is continuous on $[0, +\infty)$ because f is continuous at every point $c \in (0,+\infty)$ and is continuous at 0 from the right.

(e) $f(x) = \cos x$ is continuous on $(-\infty,+\infty)$ because f is continuous at every point $c \in (-\infty,+\infty)$.

(f) $f(x) = \tan x$ is continuous on $(0,\pi/2)$ because f is continuous at every point $c \in (0,\pi/2)$.

(g) $f(x) = \tan x$ is not continuous on $[0,\pi/2]$ because f is not continuous at $\pi/2$ from the left.

(h) $f(x) = \tan x$ is continuous on $[0,\pi/2)$ because f is continuous at every point $c \in (0,\pi/2)$ and is continuous at 0 from the right.

(i) $f(x) = 2x/(x^2 + 5)$ is continuous on $(-\infty,+\infty)$ because f is continuous at every point $c \in (-\infty,+\infty)$.

(j) $f(x) = |x - 2|/(x - 2)$ is continuous on $(-\infty,2)$ and $(2,+\infty)$ because f is continuous at every point $c \in (-\infty,2)$ and $c \in (2,+\infty)$.

(k) $f(x) = |x - 2|/(x - 2)$ is not continuous on $(-\infty,2]$ or $[2,+\infty)$ because f is not continuous at 2 from the left or from the right.

Chapter Checkout

Q&A

1. Evaluate the following

(a) $\lim\limits_{x \to 3^+} \dfrac{|x^2 - 9|}{x - 3}$

(b) $\lim\limits_{x \to 3^-} \dfrac{|x^2 - 9|}{x - 3}$

(c) $\lim\limits_{x \to 3} \dfrac{|x^2 - 9|}{x - 3}$

2. Evaluate $\lim\limits_{x \to 2^+} \dfrac{x + 2}{x^2 - 4}$

3. Evaluate $\lim\limits_{x \to +\infty} \dfrac{x^2}{x^3 - 1}$

4. Evaluate $\lim\limits_{x \to 0} \dfrac{\sin 5x}{3x}$

5. Discuss the continuity of the function

$$f(x) = \begin{cases} \dfrac{x^2 - 1}{x + 2}, & x \neq -1 \\ 2, & x = -1 \end{cases} \text{ at } x = -1.$$

Answers: 1. (a) 6 (b) –6 (c) DNE **2.** $+\infty$ **3.** 0 **4.** 5/3 **5.** f is not continuous at $x = -1$.

Chapter 3

THE DERIVATIVE

Chapter Check-In

- ❑ Understanding derivatives
- ❑ Computing basic derivatives
- ❑ Finding derivatives of more complicated functions

One of the most important applications of limits is the concept of the *derivative of a function.* In calculus, the derivative of a function is used in a wide variety of problems, and understanding it is essential to applying it to such problems.

Definition

The function f' defined by the formula

$$f'(x) = \lim_{w \to x} \frac{f(w) - f(x)}{w - x}$$

is called the *derivative* of f with respect to x. The domain of f' consists of all x in the domain of f at which the limit exists. At each such x we say the f is *differentiable.*

Figure 3-1 The derivative of a function as the limit of rise over run.

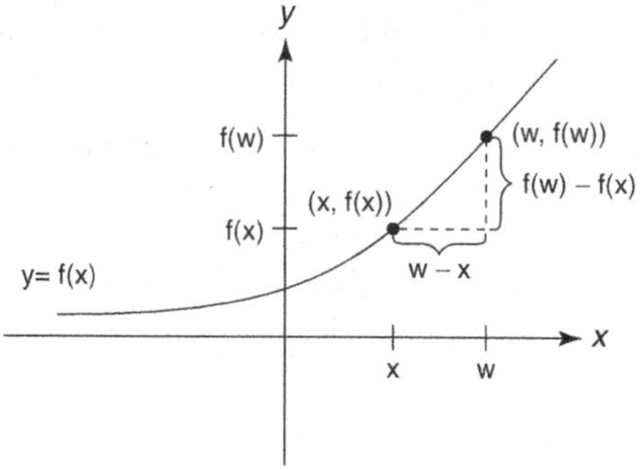

If a function is differentiable at x, then it must be continuous at x, but the converse is not necessarily true. That is, a function may be continuous at a point, but the derivative at that point may not exist. As an example, the function $f(x) = x^{1/3}$ is continuous over its entire domain or real numbers, but its derivative does not exist at $x = 0$.

Another example is the function $f(x) = |x + 2|$, which is also continuous over its entire domain of real numbers but is not differentiable at $x = -2$.

The relationship between continuity and differentiability can be summarized as follows: Differentiability implies continuity, but continuity *does not* imply differentiability.

Example 3-1: Find the derivative of $f(x) = x^2 - 5$ at the point (2,–1).

$$\frac{f(w) - f(2)}{w - 2} = \frac{(w^2 - 5) - (2^2 - 5)}{w - 2}$$

$$= \frac{w^2 - 4}{w - 2}$$

$$= \frac{(w + 2)(w - 2)}{w - 2}$$

$$\frac{f(w) - f(2)}{w - 2} = w + 2,\ w \neq 2$$

$$f'(2) = \lim_{w \to 2} (w + 2) = 2 + 2 = 4$$

Hence, the derivative of $f(x) = x^2 - 5$ at the point (2,–1) is 4.

One interpretation of the derivative of a function at a point is the **slope of the tangent line** at this point. The derivative may be thought of as the limit of the slopes of the secant lines passing through a fixed point on a curve and other points on the curve that get closer and closer to the fixed point. If this limit exists, it is defined to be the slope of the tangent line at the fixed point, $(x, f(x))$ on the graph of $y = f(x)$.

Another interpretation of the derivative is the **instantaneous velocity** of a function representing the position of a particle along a line at time t, where $y = s(t)$. The derivative may be thought of as a limit of the average velocities between a fixed time and other times that get closer and closer to the fixed time. If this limit exists, it is defined to be the instantaneous velocity at time t for the function, $y = s(t)$.

A third interpretation of the derivative is the **instantaneous rate of change** of a function at a point. The derivative may be thought of as the limit of the average rates of change between a fixed point and other points on the curve that get closer and closer to the fixed point. If this limit exists, it is defined to be the instantaneous rate of change at the fixed point $(x, f(x))$ on the graph of $y = f(x)$.

Example 3-2: Find the instantaneous velocity of $s(t) = \frac{1}{t+2}$ at the time t, and evaluate the velocity at time $t = 3$.

$$\frac{s(w) - s(t)}{w - t} = \frac{\frac{1}{w+2} - \frac{1}{t+2}}{w - t}$$

$$= \frac{1}{w-t} \cdot \frac{(t+2) - (w+2)}{(w+2)(t+2)}$$

$$= \frac{-(w-t)}{(w-t)(w+2)(t+2)}$$

$$\frac{s(w) - s(t)}{w - t} = \frac{-1}{(w+2)(t+2)}, \ w \neq t$$

$$s'(t) = \lim_{w \to t} \frac{-1}{(w+2)(t+2)}$$

$$= \frac{-1}{(t+2)^2}$$

$$s'(3) = \frac{-1}{(3+2)^2} = \frac{-1}{5^2} = \frac{-1}{25}$$

Hence, the instantaneous velocity of $s(t) = 1/(t + 2)$ at time $t = 3$ is $-1/25$. The negative velocity indicates that the particle is moving in the negative direction.

A number of different notations are used to represent the derivative of a function $y = f(x)$ with $f'(x)$ being most common. Some others are y', dy/dx, df/dx, $df(x)/dx$, $D_x f$, and $D_x f(x)$, and you should be able to use any of these in selected problems.

Differentiation Rules

Many differentiation rules can be proven using the limit definition of the derivative and are also useful in finding the derivatives of applicable functions. To eliminate the need of using the formal definition for every application of the derivative, some of the more useful formulas are listed here.

(1) If $f(x) = c$, where c is a constant, the $f'(x) = 0$.

(2) If $f(x) = c \cdot g(x)$, then $f'(x) = c \cdot g'(x)$.

(3) Sum Rule: If $f(x) = g(x) + h(x)$, then $f'(x) = g'(x) + h'(x)$.

(4) Difference Rule: If $f(x) = g(x) - h(x)$, then $f'(x) = g'(x) - h'(x)$.

(5) Product Rule: If $f(x) = g(x) \cdot h(x)$, then $f'(x) = g(x) \cdot h'(x) + h(x) \cdot g'(x)$.

(6) Quotient Rule: If $f(x) = \frac{g(x)}{h(x)}$ and $h(x) \neq 0$, then

$$f'(x) = \frac{h(x) \cdot g'(x) - g(x) \cdot h'(x)}{[h(x)]^2}$$

(7) Power Rule: If $f(x) = x^n$, then $f'(x) = nx^{n-1}$.

Example 3-3: Find $f'(x)$ if $f(x) = 6x^3 + 5x^2 + 9$.

$$\begin{aligned} f'(x) &= 6 \cdot 3x^2 + 5 \cdot 2x^1 + 0 \\ &= 18x^2 + 10x \end{aligned}$$

Example 3-4: Find y' if $y = (3x + 4)(2x^2 - 3x + 5)$.

$$\begin{aligned} y' &= (3x + 4) + (4x - 3) + (2x^2 - 3x + 5)(3) \\ &= 12x^2 + 7x - 12 + 6x^2 - 9x + 15 \\ &= 18x^2 - 2x + 3 \end{aligned}$$

Example 3-5: Find $\frac{dy}{dx}$ if $y = \frac{3x+5}{2x-3}$.

$$\frac{dy}{dx} = \frac{(2x-3)(3)-(3x+5)(2)}{(2x-3)^2}$$
$$= \frac{6x-9-6x-10}{(2x-3)^2}$$
$$= \frac{-19}{(2x-3)^2}$$

Example 3-6: Find $f'(x)$ if $f(x) = x^5 - \sqrt{x} + \frac{1}{x^3}$.

$$\text{Because } f(x) = x^5 - x^{1/2} + x^{-3}$$
$$f'(x) = 5x^4 - \frac{1}{2}x^{-1/2} - 3x^{-4}$$
$$= 5x^4 - \frac{1}{2\sqrt{x}} - \frac{3}{x^4}$$

Example 3-7: Find $f'(3)$ if $f(x) = x^2 - 8x + 3$.

$$f'(x) = 2x - 8$$
$$f'(3) = (2)(3) - 8$$
$$= -2$$

Example 3-8: If $y = \frac{4}{x+2}$, find y' at (2, 1).

$$y' = \frac{(x+2)(0)-4(1)}{(x+2)^2}$$
$$= \frac{-4}{(x+2)^2}$$
$$\text{At } (2, 1),\quad y' = \frac{-4}{(2+2)^2}$$
$$= \frac{-4}{16}$$
$$= -\frac{1}{4}$$

Example 3-9: Find the slope of the tangent line to the curve $y = 12 - 3x^2$ at the point $(-1,9)$.

Because the slope of the tangent line to a curve is the derivative, you find that $y' = -6x$; hence, at $(-1,9)$, $y' = 6$, and the tangent line has slope 6 at the point $(-1,9)$.

Trigonometric Function Differentiation

The six trigonometric functions also have differentiation formulas that can be used in application problems of the derivative. The rules are summarized as follows:

(1) If $f(x) = \sin x$, then $f'(x) = \cos x$.
(2) If $f(x) = \cos x$, then $f'(x) = -\sin x$.
(3) If $f(x) = \tan x$, then $f'(x) = \sec^2 x$.
(4) If $f(x) = \cot x$, then $f'(x) = -\csc^2 x$.
(5) If $f(x) = \sec x$, then $f'(x) = \sec x \tan x$.
(6) If $f(x) = \csc x$, then $f'(x) = -\csc x \cot x$.

Note that rules (3) to (6) can be proven using the quotient rule along with the given function expressed in terms of the sine and cosine functions, as illustrated in the following example.

Example 3-10: Use the definition of the tangent function and the quotient rule to prove if $f(x) = \tan x$, than $f'(x) = \sec^2 x$.

$$f(x) = \tan x$$

$$= \frac{\sin x}{\cos x}$$

$$f'(x) = \frac{\cos x \cdot \cos x - (\sin x)(-\sin x)}{\cos^2 x}$$

$$= \frac{\cos^2 x + \sin^2 x}{\cos^2 x}$$

$$= \frac{1}{\cos^2 x}$$

$$= \sec^2 x$$

Example 3-11: Find y' if $y = x^3 \cot x$.

$$y' = x^3(-\csc^2 x) + (\cot x)(3x^2)$$
$$= 3x^2 \cot x - x^3 \csc^2 x$$

Example 3-12: Find $f'\left(\frac{\pi}{4}\right)$ if $f(x) = 5\sin x + \cos x$.

$$f'(x) = 5\cos x - \sin x$$
$$f'\left(\frac{\pi}{4}\right) = 5\cos\frac{\pi}{4} - \sin\frac{\pi}{4}$$
$$= \frac{5\sqrt{2}}{2} - \frac{\sqrt{2}}{2}$$
$$= \frac{4\sqrt{2}}{2}$$
$$= 2\sqrt{2}$$

Example 3-13: Find the slope of the tangent line to the curve $y = \sin x$ at the point $(\pi/2, 1)$

Because the slope of the tangent line to a curve is the derivative, you find that $y' = \cos x$; hence, at $(\pi/2, 1)$, $y' = \cos \pi/2 = 0$, and the tangent line has a slope 0 at the point $(\pi/2, 1)$. Note that the geometric interpretation of this result is that the tangent line is horizontal at this point on the graph of $y = \sin x$.

Chain Rule

The **chain rule** provides us a technique for finding the derivative of composite functions, with the number of functions that make up the composition determining how many differentiation steps are necessary. For example, if a composite function $h(x)$ is defined as

$$h(x) = (f \circ g)(x) = f[g(x)]$$
$$\text{then } h'(x) = f'[g(x)] \cdot g'(x)$$

Note that because two functions, f and g, make up the composite function h, you have to consider the derivatives f' and g' in differentiating $h(x)$.

If a composite function $r(x)$ is defined as

$$r(x) = (m \circ n \circ p)(x) = m\{n[p(x)]\}$$
$$\text{then } r'(x) = m'\{n[p(x)]\} \cdot n'[p(x)] \cdot p'(x)$$

Here, three functions—m, n, and p—make up the composition function r; hence, you have to consider the derivatives m', n', and p' in differentiating $r(x)$. A technique that is sometimes suggested for differentiating composite functions is to work from the "outside to the inside" functions to establish a sequence for each of the derivatives that must be taken.

Example 3-14: Find $h'(x)$ if $h(x) = (3x^2 + 5x - 2)^8$.

$$\begin{aligned} h'(x) &= 8(3x^2 + 5x - 2)^7 \cdot (6x + 5) \\ &= 8(6x + 5)(3x^2 + 5x - 2)^7 \end{aligned}$$

Example 3-15: Find $f'(x)$ if $f(x) = \tan(\sec x)$.

$$\begin{aligned} f'(x) &= \sec^2(\sec x) \cdot \sec x \tan x \\ &= \sec x \tan x \sec^2(\sec x) \end{aligned}$$

Example 3-16: Find $\frac{dy}{dx}$ if $y = \sin^3(3x - 1)$.

$$\begin{aligned} \frac{dy}{dx} &= 3\sin^2(3x - 1) \cdot \cos(3x - 1) \cdot (3) \\ &= 9\cos(3x - 1)\sin^2(3x - 1) \end{aligned}$$

Example 3-17: Find $f'(2)$ if $f(x) = \sqrt{5x^2 + 3x - 1}$.

Because $f(x) = (5x^2 + 3x - 1)^{1/2}$

$$\begin{aligned} f'(x) &= \frac{1}{2}(5x^2 + 3x - 1)^{-1/2}(10x + 3) \\ &= \frac{10x + 3}{2\sqrt{5x^2 + 3x - 1}} \\ f'(2) &= \frac{10 \cdot 2 + 3}{2\sqrt{5(2)^2 + 3 \cdot 2 - 1}} \\ &= \frac{23}{2\sqrt{25}} \\ &= \frac{23}{10} \end{aligned}$$

Example 3-18: Find the slope of the tangent line to a curve $y = (x^2 - 3)^5$ at the point (–1, –32).

Because the slope of the tangent line to a curve is the derivative, you find that

$$\begin{aligned} y' &= 5(x^2-3)^4(2x) \\ &= 10x(x^2-3)^4 \\ \text{hence, at } (-1,-32) \quad y' &= 10(-1)[(-1)^2-3]^4 \\ &= (-10)(-2)^4 \\ &= -160 \end{aligned}$$

which represents the slope of the tangent line at the point (–1,–32).

Implicit Differentiation

In mathematics, some equations in x and y do not explicitly define y as a function x and cannot be easily manipulated to solve for y in terms of x, even though such a function may exist. When this occurs, it is implied that there exists a function $y = f(x)$ such that the given equation is satisfied. The technique of **implicit differentiation** allows you to find the derivative of y with respect to x without having to solve the given equation for y. The chain rule must be used whenever the function y is being differentiated because of our assumption that y may be expressed as a function of x.

Example 3-19: Find $\frac{dy}{dx}$ if $x^2y^3 - xy = 10$.

Differentiating implicitly with respect to x, you find that

$$x^2 \cdot 3y^2 \cdot \frac{dy}{dx} + y^3 \cdot 2x - x \cdot \frac{dy}{dx} - y = 0$$

$$3x^2y^2\frac{dy}{dx} - x\frac{dy}{dx} = y - 2xy^3$$

$$(3x^2y^2 - x)\frac{dy}{dx} = y - 2xy^3$$

$$\frac{dy}{dx} = \frac{y - 2xy^3}{3x^2y^2 - x}$$

$$\text{or} \quad \frac{dy}{dx} = \frac{2xy^3 - y}{x - 3x^2y^2}$$

Example 3-20: Find y' if $y = \sin x + \cos y$.

Differentiating implicitly with respect to x, you find that

$$y' = \cos x - \sin y \cdot y'$$
$$1 \cdot y' + \sin y \cdot y' = \cos x$$
$$y'(1 + \sin y) = \cos x$$
$$y' = \frac{\cos x}{1 + \sin y}$$

Example 3-21: Find y' at $(-1,1)$ if $x^2 + 3xy + y^2 = -1$.

Differentiating implicitly with respect to x, you find that

$$2x + 3x \cdot y' + 3y + 2y \cdot y' = 0$$
$$3x \cdot y' + 2y \cdot y' = -2x - 3y$$
$$y'(3x + 2y) = -2x - 3y$$
$$y' = \frac{-2x - 3y}{3x + 2y}$$
$$\text{At the point}(-1, 1), y' = \frac{(-2)(-1) - 3(1)}{3(-1) + 2(1)}$$
$$= \frac{-1}{-1}$$
$$= 1$$

Example 3-22: Find the slope of the tangent line to the curve $x^2 + y^2 = 25$ at the point $(3,-4)$.

Because the slope of the tangent line to a curve is the derivative, differentiate implicitly with respect to x, which yields

$$2x + 2y \cdot y' = 0$$
$$2y \cdot y' = -2x$$
$$y' = \frac{-2x}{2y}$$
$$= \frac{-x}{y}$$

hence, at $(3,-4)$, $y' = -3/-4 = 3/4$, and the tangent line has slope $3/4$ at the point $(3,-4)$.

Higher Order Derivatives

Because the derivative of a function $y = f(x)$ is itself a function $y' = f'(x)$, you can take the derivative of $f'(x)$, which is generally referred to as the *second derivative of f(x)* and written $f''(x)$ or $f^{(2)}(x)$. This differentiation process can be continued to find the third, fourth, and successive derivatives of $f(x)$, which are called **higher order derivatives** of $f(x)$. Because the "prime" notation for derivatives would eventually become somewhat messy, it is preferable to use the numerical notation $f^{(n)}(x) = y^{(n)}$ to denote the *n*th derivative of $f(x)$. Chapter 4 provides some applications of the second derivative in curve sketching and in distance, velocity, and acceleration problems.

Example 3-23: Find the first, second, and third derivatives of $f(x) = 5x^4 - 3x^3 + 7x^2 - 9x + 2$.

$$f'(x) = 20x^3 - 9x^2 + 14x - 9$$
$$f''(x) = f^{(2)}(x) = 60x^2 - 18x + 14$$
$$f'''(x) = f^{(3)}(x) = 120x - 18$$

Example 3-24: Find the first, second, and third derivatives of $y = \sin^2 x$.

$$y' = 2\sin x\cos x$$
$$y'' = 2\cos x\cos x + 2\sin x(-\sin x)$$
$$= 2\cos^2 x - 2\sin^2 x$$
$$y''' = 2\cdot 2\cos x(-\sin x) - 2\cdot 2\sin x\cos x$$
$$= -4\sin x\cos x - 4\sin x\cos x$$
$$= -8\sin x\cos x$$

Example 3-25: Find $f^{(3)}(4)$ if $f(x) = \sqrt{x}$.

$$\text{Because } f(x) = \sqrt{x} = x^{1/2}$$
$$f'(x) = \frac{1}{2}x^{-1/2}$$
$$f''(x) = -\frac{1}{4}x^{-3/2}$$
$$f'''(x) = \frac{3}{8}x^{-5/2}$$
$$\text{hence, } f'''(4) = \frac{3}{8}(4)^{-5/2}$$
$$= \frac{3}{8}\left(\frac{1}{32}\right)$$
$$= \frac{2}{256}$$

Differentiation of Inverse Trigonometric Functions

Each of the six basic trigonometric functions have corresponding inverse functions when appropriate restrictions are placed on the domain of the original functions. All the inverse trigonometric functions have derivatives, which are summarized as follows:

(1) If $f(x) = \sin^{-1} x = \arcsin x, -\frac{\pi}{2} \leq f(x) \leq \frac{\pi}{2}$ then

$$f'(x) = \frac{1}{\sqrt{1-x^2}}.$$

(2) If $f(x) = \cos^{-1} x = \arccos x, 0 \leq f(x) \leq \pi$ then

$$f'(x) = \frac{-1}{\sqrt{1-x^2}}.$$

(3) If $f(x) = \tan^{-1} x = \arctan x, -\frac{\pi}{2} < f(x) < \frac{\pi}{2}$ then

$$f'(x) = \frac{1}{1+x^2}.$$

(4) If $f(x) = \cot^{-1} x = \operatorname{arccot} x, 0 < f(x) < \pi$ then

$$f'(x) = \frac{-1}{1+x^2}.$$

(5) If $f(x) = \sec^{-1} x = \operatorname{arcsec} x, 0 \leq f(x) \leq \pi, f(x) \neq \frac{\pi}{2}$ then

$$f'(x) = \frac{1}{|x|\sqrt{x^2-1}}.$$

(6) If $f(x) = \csc^{-1} x = \operatorname{arccsc} x, -\frac{\pi}{2} \leq f(x) \leq \frac{\pi}{2}, f(x) \neq 0$

then $f'(x) = \dfrac{-1}{|x|\sqrt{x^2-1}}$.

Example 3-26: Find $f'(x)$ if $f(x) = \cos^{-1}(5x)$.

$$f'(x) = \frac{-1}{\sqrt{1-(5x)^2}} \cdot 5$$

$$f'(x) = \frac{-5}{\sqrt{1-25x^2}}$$

Example 3-27: Find y' if $y = \arctan\left(\sqrt{x^3}\right)$.

$$\text{Because } y = \arctan(x^{3/2})$$

$$y' = \frac{1}{1 + (x^{3/2})^2} \cdot \frac{3}{2} x^{1/2}$$

$$= \frac{1}{1 + x^3} \cdot \frac{3}{2} x^{1/2}$$

$$y' = \frac{3\sqrt{x}}{2(1 + x^3)}$$

Differentiation of Exponential and Logarithmic Functions

Exponential functions and their corresponding inverse functions, called *logarithmic functions,* have the following differentiation formulas:

(1) If $f(x) = e^x$, then $f'(x) = e^x$.

(2) If $f(x) = a^x, a > 0, a \neq 1$, then $f'(x) = (\ln a) \cdot a^x$.

(3) If $f(x) = \ln x$, then $f'(x) = \frac{1}{x}$.

(4) If $f(x) = \log_a x, a > 0, a \neq 1$, then $f'(x) = \frac{1}{(\ln a) \cdot x}$.

Note that the exponential function $f(x) = e^x$ has the special property that its derivative is the function itself, $f'(x) = e^x = f(x)$.

Example 3-28: Find $f'(x)$ if $f(x) = e^{x^2 + 5}$.

$$f'(x) = e^{x^2 + 5} \cdot 2x$$

$$f'(x) = 2x \cdot e^{x^2 + 5}$$

Example 3-29: Find y' if $y = 5^{\sqrt{x}}$.

$$y' = (\ln 5) \cdot 5^{\sqrt{x}} \cdot \frac{1}{2} x^{-1/2}$$

$$= (\ln 5) \cdot 5^{\sqrt{x}} \cdot \frac{1}{2\sqrt{x}}$$

$$y' = \frac{(\ln 5) \cdot 5^{\sqrt{x}}}{2\sqrt{x}}$$

Example 3-30: Find $f'(x)$ if $f(x) = \ln(\sin x)$.

$$f'(x) = \frac{1}{\sin x} \cdot \cos x$$

$$= \frac{\cos x}{\sin x}$$

$$f'(x) = \cot x$$

Example 3-31: Find $\frac{dy}{dx}$ if $y = \log_{10}(4x^2 - 3x - 5)$.

$$\frac{dy}{dx} = \frac{1}{(\ln 10)(4x^2 - 3x - 5)} \cdot (8x - 3)$$

$$\frac{dy}{dx} = \frac{8x - 3}{(\ln 10)(4x^2 - 3x - 5)}$$

Chapter Checkout

Q&A

1. Find $f'(x)$ if $f(x) = 5x^3 - 7 + x^2 \sin x + e^x/x$.
2. Find dy/dx if $y = \sqrt{x^4 - 2x + 1}$
3. Find y' if $y = \ln(\cos x) - e^{5-x}$.
4. Find $f''(1/2)$ if $f(x) = \sin^{-1} x$.
5. Find the slope of the tangent line to the curve $x^2 + xy + y^2 = 4$ at the point $(-2,2)$.

Answers: 1. $15x^2 + x^2 \cos x + 2x \sin x + (xe^x - e^x)/x^2$ **2.** $dy/dx = (2x^3 - 1)/(x^4 - 2x + 1)^{1/2}$ **3.** $e^{5-x} - \tan x$ **4.** $\frac{2\sqrt{3}}{3}$ **5.** 1

Chapter 4

APPLICATIONS OF THE DERIVATIVE

Chapter Check-In

- ❑ Using the derivative to understand the graph of a function
- ❑ Locating maximum and minimum values of a function
- ❑ Finding velocity and acceleration
- ❑ Relating rates of change
- ❑ Approximating quantities by using derivatives

The derivative of a function has many applications to problems in calculus. It may be used in curve sketching; solving maximum and minimum problems; solving distance; velocity, and acceleration problems; solving related rate problems; and approximating function values.

Tangent and Normal Lines

As previously noted, the derivative of a function at a point, is the slope of the tangent line at this point. The **normal line** is defined as the line that is perpendicular to the tangent line at the point of tangency. Because the slopes of perpendicular lines (neither of which is vertical) are negative reciprocals of one another, the slope of the normal line to the graph of $f(x)$ is $-1/f'(x)$.

Example 4-1: Find the equation of the tangent line to the graph of $f(x) = \sqrt{x^2 + 3}$ at the point (–1,2).

$$f(x) = (x^2 + 3)^{1/2}$$

$$f'(x) = \frac{1}{2}(x^2 + 3)^{-1/2} \cdot (2x)$$

$$f'(x) = \frac{x}{\sqrt{x^2 + 3}}$$

At the point (–1,2), $f'(-1) = -1/2$ and the equation of the line is

$$y - y_1 = m(x - x_1)$$

$$y - 2 = -\frac{1}{2}(x + 1)$$

$$2y - 4 = -x - 1$$

$$x + 2y - 3 = 0$$

Example 4-2: Find the equation of the normal line to the graph of $f(x) = \sqrt{x^2 + 3}$ at the point (–1, 2).

From Example 4-1, you find that $f'(-1) = -1/2$ and the slope of the normal line is $-1/f'(-1) = 2$; hence, the equation of the normal line at the point (–1,2) is

$$y - y_1 = m(x - x_1)$$

$$y - 2 = 2(x + 1)$$

$$y - 2 = 2x + 2$$

$$2x - y + 4 = 0$$

Critical Points 4.2/ET5.2

Points on the graph of a function where the derivative is zero or the derivative does not exist are important to consider in many application problems of the derivative. The point $(x, f(x))$ is called a **critical point** of $f(x)$ if x is in the domain of the function and either $f'(x) = 0$ or $f'(x)$ does not exist. If $(x, f(x))$ is a critical point of $f(x)$, then x is called a **critical number** of $f(x)$. The geometric interpretation of what is taking place at a critical point is that the tangent line is either horizontal, vertical, or does not exist at that point on the curve.

Example 4-3: Find all critical points of $f(x) = x^4 - 8x^2$.

Because $f(x)$ is a polynomial function, its domain is all real numbers.

$$f'(x) = 4x^3 - 16x$$

$$f'(x) = 0 \Rightarrow 4x^3 - 16x = 0$$

$$4x(x^2 - 4) = 0$$

$$4x(x + 2)(x - 2) = 0$$

$$x = 0, x = -2, x = 2$$

$$f(-2) = (-2)^4 - 8(-2)^2 = -16$$

$$f(0) = (0)^4 - 8(0)^2 = 0$$

$$f(2) = (2)^4 - 8(2)^2 = -16$$

Hence, the critical points of $f(x)$ are (–2,–16), (0,0), and (2,–16), and the critical numbers are x = –2, 0, and 2.

Example 4-4: Find all critical points of $f(x) = \sin x + \cos x$ on $[0,2\pi]$.

The domain of $f(x)$ is restricted to the closed interval $[0,2\pi]$.

$$f'(x) = \cos x - \sin x$$

$$f'(x) = 0 \Rightarrow \cos x - \sin x = 0$$

$$\cos x = \sin x$$

$$x = \frac{\pi}{4}, \frac{5\pi}{4}$$

$$f\left(\frac{\pi}{4}\right) = \sin\frac{\pi}{4} + \cos\frac{\pi}{4} = \frac{\sqrt{2}}{2} + \frac{\sqrt{2}}{2} = \sqrt{2}$$

$$f\left(\frac{5\pi}{4}\right) = \sin\frac{5\pi}{4} + \cos\frac{5\pi}{4} = \frac{-\sqrt{2}}{2} + \frac{-\sqrt{2}}{2} = -\sqrt{2}$$

hence, the critical points of $f(x)$ are $(\pi/4, \sqrt{2})$ and $(5\pi/4, -\sqrt{2})$.

Extreme Value Theorem

An important application of critical points is in determining possible maximum and minimum values of a function on certain intervals. The **Extreme Value Theorem** guarantees both a maximum and minimum value for a function under certain conditions. It states the following:

If a function $f(x)$ is continuous on a closed interval $[a,b]$, then $f(x)$ has both a maximum and minimum value on $[a,b]$.

The procedure for applying the Extreme Value Theorem is to first establish that the function is continuous on the closed interval. The next step is to determine all critical points in the given interval and evaluate the function at these critical points and at the endpoints of the interval. The largest function value from the previous step is the maximum value, and the smallest function value is the minimum value of the function on the given interval.

Example 4-5: Find the maximum and minimum values of $f(x) = \sin x + \cos x$ on $[0,2\pi]$.

The function is continuous on $[0,2\pi]$, and from Example 4-4, the critcal points are $\left(\pi/4, \sqrt{2}\right)$ and $\left(5\pi/4, -\sqrt{2}\right)$. The function values at the end points of the interval are $f(0)=1$ and $f(2\pi)=1$; hence, the maximum function value of $f(x)$ is $\sqrt{2}$ at $x=\pi/4$, and the minimum function value of $f(x)$ is $-\sqrt{2}$ at $x = 5\pi/4$.

Note that for this example the maximum and minimum both occur at critical points of the function.

Example 4-6: Find the maximum and minimum values of $f(x) = x^4 - 3x^3 - 1$ on $[-2,2]$.

The function is continuous on $[-2,2]$, and its derivative is $f'(x) = 4x^3 - 9x^2$.

$$f'(x) = 0 \Rightarrow 4x^3 - 9x^2 = 0$$

$$x^2(4x - 9) = 0$$

$$x = 0, x = \frac{9}{4}$$

Because $x = 9/4$ is not in the interval $[-2,2]$, the only critical point occurs at $x = 0$ which is $(0,-1)$. The function values at the endpoints of the interval are $f(2) = -9$ and $f(-2) = 39$; hence, the maximum function value 39 at $x = -2$, and the minimum function value is -9 at $x = 2$. Note the importance of the closed interval in determining which values to consider for critical points.

Mean Value Theorem

The **Mean Value Theorem** establishes a relationship between the slope of a tangent line to a curve and the secant line through points on a curve at the endpoints of an interval. The theorem is stated as follows.

If a function $f(x)$ is continuous on a closed interval [a,b] and differentiable on an open interval (a,b), then at least one number $c \in (a,b)$ exists such that

$$f'(c) = \frac{f(b) - f(a)}{b - a}$$

Figure 4-1 The Mean Value Theorem.

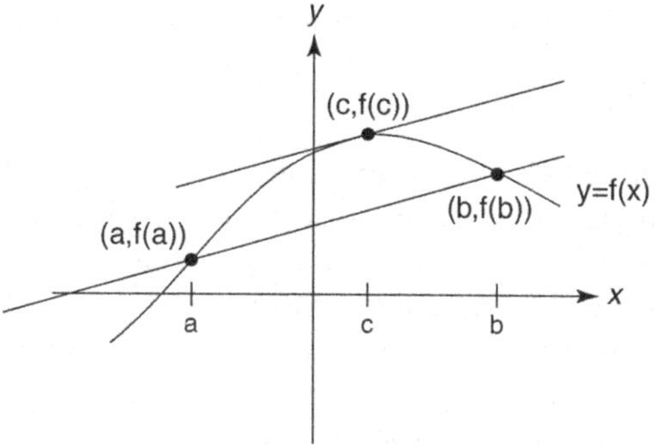

Geometrically, this means that the slope of the tangent line will be equal to the slope of the secant line through $(a,f(a))$ and $(b,f(b))$ for at least one point on the curve between the two endpoints. Note that for the special case where $f(a) = f(b)$, the theorem guarantees at least one critical point, where $f'(c) = 0$ on the open interval (a,b).

Example 4-7: Verify the conclusion of the Mean Value Theorem for $f(x) = x^2 - 3x - 2$ on [–2,3].

The function is continuous on [–2,3] and differentiable on (–2,3). The slope of the secant line through the endpoint values is

$$\frac{f(3) - f(-2)}{3 - (-2)} = \frac{-2 - 8}{5} = \frac{-10}{5} = -2$$

The slope of the tangent line is

$$f'(x) = 2x - 3$$

$$f'(x) = -2 \Rightarrow 2x - 3 = -2$$

$$2x = 1$$

$$x = \frac{1}{2}$$

Because $1/2 \in [-2,3]$, the c value referred to in the conclusion of the Mean Value Theorem is $c = 1/2$

Increasing/Decreasing Functions 4.1/ET5.1

If $f(x_1) < f(x_2)$ for any points x_1, x_2 in an interval I with $x_1 < x_2$, then $f(x)$ is said to be **increasing** *on I.* If $f(x_1) > f(x_2)$ for any points x_1, x_2 in an interval I with $x_1 < x_2$, then $f(x)$ is said to be **decreasing** *on I.* The derivative of $f(x)$ may be used to determine whether the function is increasing or decreasing on intervals in its domain. Note that if $f'(x) > 0$ at each point in I then $f(x)$ is increasing on I, and if $f'(x) < 0$ at each point in I then $f(x)$ is decreasing on I. Because the derivative is zero or does not exist only at critical points of the function, it must be positive or negative at all other points where the function is defined.

In determining intervals where a function is increasing or decreasing, you first find domain values where all critical points will occur; then, test all intervals in the domain of the function to the left and to the right of these values to determine if the derivative is positive or negative. If $f'(x) > 0$, then f is increasing on the interval, and if $f'(x) < 0$, then f is decreasing on the interval. This and other information may be used to show a reasonably accurate sketch of the graph of the function.

Example 4-8: For $f(x) = x^4 - 8x^2$ determine all intervals where f is increasing or decreasing.

As noted in Example 4-3, the domain of $f(x)$ is all real numbers, and its critical points occur at $x = -2$, 0, and 2. Testing all intervals to the left and right of these values for $f'(x) = 4x^3 - 16x$, you find that

$$f'(x) < 0 \text{ on } (-\infty, -2)$$
$$f'(x) > 0 \text{ on } (-2, 0)$$
$$f'(x) < 0 \text{ on } (0, 2)$$
$$f'(x) > 0 \text{ on } (2, +\infty)$$

hence, f is increasing on $(-2,0)$ and $(2,+\infty)$ and decreasing on $(-\infty,-2)$ and $(0,2)$.

Example 4-9: For $f(x) = \sin x + \cos x$ on $[0,2\pi]$, determine all intervals where f is increasing or decreasing.

As noted in Example 4-4, the domain of $f(x)$ is restricted to the closed interval $[0,2\pi]$, and its critical numbers are $\pi/4$ and $5\pi/4$. Testing all intervals to the left and right of these values for $f'(x) = \cos x - \sin x$, you find that

$$f'(x) > 0 \text{ on} \left[0, \frac{\pi}{4}\right)$$

$$f'(x) < 0 \text{ on} \left(\frac{\pi}{4}, \frac{5\pi}{4}\right)$$

$$f'(x) > 0 \text{ on} \left(\frac{5\pi}{4}, 2\pi\right]$$

hence, f is increasing on $[0, \pi/4)$ and $(5\pi/4, 2\pi]$ and decreasing on $(\pi/4, 5\pi/4)$.

First Derivative Test for Relative Extrema

If the derivative of a function changes sign around a critical point, the function is said to have a *realtive* (or *local*) *extremum* at that point. If the derivative changes from positive (increasing function) to negative (decreasing function), the function has a *realtive* (or *local*) *maximum* at the critical point. If, however, the derivative changes from negative (decreasing function) to positive (increasing function), the function has a *realtive (local) minimum* at the critical point. When this technique is used to determine relative maximum or minimum function values, it is called the **First Derivative Test for Local Extrema.** Note that there is no guarantee that the derivative will change signs, and therefore, it is essential to test each interval around a critical number.

Example 4-10: If $f(x) = x^4 - 8x^2$, determine all relative extrema for the function.

As noted in Example 4-8, $f(x)$ critical numbers are $x = -2, 0, 2$. Because $f'(x)$ changes from negative to positive around -2 and 2, f has a relative minimum at $(-2,-16)$ and $(2,-16)$. Also, $f'(x)$ changes from positive to negative around 0, and hence, f has a relative maximum at $(0,0)$.

Example 4-11: If $f(x) = \sin x + \cos x$ on $[0,2\pi]$, determine all relative extrema for the function.

As noted in Example 4-9, $f(x)$ critical numbers are $x = \pi/4$ and $5\pi/4$. Because $f'(x)$ changes from positive to negative around $\pi/4$, f has a relative maximum at $(\pi/4, \sqrt{2})$. Also $f'(x)$ changes from negative to positive around $5\pi/4$, and hence, f has a relative minimum at $(5\pi/4, -\sqrt{2})$.

Second Derivative Test for Relative Extrema

The second derivative may be used to determine local extrema of a function under certain conditions. If a function has a critical number for which $f'(x) = 0$ and the second derivative is positive at this point, then f has a local minimum here. If, however, the function has a critical number for which $f'(x) = 0$ and the second derivative is negative at this point, then f has local maximum here. This technique is called **Second Derivative Test for Relative Extrema.**

Three possible situations could occur that would rule out the use of the Second Derivative Test for Relative Extrema:

$$(1)\quad f'(x) = 0 \text{ and } f''(x) = 0$$

$$(2)\quad f'(x) = 0 \text{ and } f''(x) \text{ does not exist}$$

$$(3)\quad f'(x) \text{ does not exist}$$

Under any of these conditions, the First Derivative Test would have to be used to determine any relative extrema. Another drawback to the Second Derivative Test is that for some functions, the second derivative is difficult or tedious to find. As with the previous situations, revert back to the First Derivative Test to determine any relative extrema.

Example 4-12: Find any relative extrema of $f(x) = x^4 - 8x^2$ using the Second Derivative Test.

As noted in Example 4-3, $f'(x) = 0$ at $x = -2$, 0, and 2. Because $f''(x) = 12x^2 - 16$, you find that $f''(-2) = 32 > 0$, and f has a relative minimum at $(-2,-16)$; $f''(0) = -16 < 0$, and f has relative maximum at $(0,0)$; and $f''(2) = 32 > 0$, and f has a relative minimum $(2,-16)$. These results agree with the relative extrema determined in Example 4-10 using the First Derivative Test on $f(x) = x^4 - 8x^2$.

Example 4-13: Find any relative extrema of $f(x) = \sin x + \cos x$ on $[0,2\pi]$ using the Second Derivative Test.

As noted in Example 4–4, $f'(x) = 0$ at $x = \pi/4$ and $5\pi/4$. Because $f''(x) = -\sin x - \cos x$, you find that $f''(\pi/4) = -\sqrt{2}$ and f has a relative maximum at $(\pi/4, \sqrt{2})$. Also, $f''(5\pi/4) = \sqrt{2}$, and f has a relative minimum at $(5\pi/4, -\sqrt{2})$. These results agree with the relative extrema determined in Example 4-11 using the First Derivative Test on $f(x) = -\sin x - \cos x$ on $[0,2\pi]$.

Concavity and Points of Inflection

The second derivative of a function may also be used to determine the general shape of its graph on selected intervals. A function $f(x)$ is said to be **concave up** *on an interval I* if $f'(x)$ is increasing on I; $f(x)$ is **concave down** *on I* if $f'(x)$ is decreasing on I. Note that if $f''(x) > 0$ at each point in I then $f(x)$ is concave up on I, and if $f''(x) < 0$ at each point in I then $f(x)$ is concave down on I. If $f(x)$ changes from concave up to concave down or vice versa around a point on its graph, then that point is called a **point of inflection** of $f(x)$.

In determining intervals where a function is concave up or concave down, you first find domain values where $f''(x) = 0$ or $f''(x)$ does not exist. Then test all intervals around these values in the second derivative of the function. If $f''(x)$ changes sign, then $(x, f(x))$ is a point of inflection of the function. As with the First Derivative Test for Relative Extrema, there is no guarantee that the second derivative will change signs, and therefore, it is essential to test each interval around the values for which $f''(x) = 0$ or does not exist.

Geometrically, a function is concave up on an interval if its graph behaves like a portion of a parabola that opens upward. Likewise, a function that is concave down on an interval looks like a portion of a parabola that opens downward. If the graph of a function is linear on some interval in its domain, its second derivative will be zero, and it is said to have no concavity on that interval.

Example 4-14: Determine the concavity of $f(x) = x^3 - 6x^2 - 12x + 2$ and identify any points of inflection of $f(x)$.

Because $f(x)$ is a polynomial function, its domain is all real numbers.

$$f'(x) = 3x^2 - 12x - 12$$
$$f''(x) = 6x - 12$$
$$f''(x) = 0 \Rightarrow 6x - 12 = 0$$
$$6x = 12$$
$$x = 2$$

Testing the intervals to the left and right of $x = 2$ for $f''(x) = 6x - 12$, you find that

$$f''(x) < 0 \text{ on } (-\infty, 2)$$
$$\text{and } f''(x) > 0 \text{ on } (2, +\infty)$$

hence, f is concave down on $(-\infty, 2)$ and concave up on $(2, +\infty)$, and function has a point of inflection at $(2, -38)$

Example 4-15: Determine the concavity of $f(x) = \sin x + \cos x$ on $[0,2\pi]$ and identify any points of inflection of $f(x)$.

The domain of $f(x)$ is restricted to the closed interval $[0,2\pi]$.

$$f'(x) = \cos x - \sin x$$

$$f''(x) = -\sin x - \cos x$$

$$f''(x) = 0 \Rightarrow -\sin x - \cos x = 0$$

$$-\sin x = \cos x$$

$$x = \frac{3\pi}{4}, \frac{7\pi}{4}$$

Testing all intervals to the left and right of these values for $f''(x) = -\sin x - \cos x$, you find that

$$f''(x) < 0 \text{ on } \left[0, \frac{3\pi}{4}\right]$$

$$f''(x) > 0 \text{ on } \left(\frac{3\pi}{4}, \frac{7\pi}{4}\right)$$

$$f''(x) < 0 \text{ on } \left(\frac{7\pi}{4}, 2\pi\right)$$

hence, f is concave down on $[0,3\pi/4)$ and $(7\pi/4,2\pi]$ and concave up on $(3\pi/4,7\pi/4)$ and has points of inflection at $(3\pi/4,0)$ and $(7\pi/4,0)$.

Maximum/Minimum Problems

Many application problems in calculus involve functions for which you want to find maximum or minimum values. The restrictions stated or implied for such functions will determine the domain from which you must work. The function, together with its domain, will suggest which technique is appropriate to use in determining a maximum or minimum value—the Extreme Value Theorem, the First Derivative Test, or the Second Derivative Test.

Example 4-16: A rectangular box with a square base and no top is to have a volume of 108 cubic inches. Find the dimensions for the box that require the least amount of material.

The function that is to be minimized is the surface area (S) while the volume (V) remains fixed at 108 cubic inches (Figure 4-2).

Figure 4-2 The open-topped box for Example 4-16.

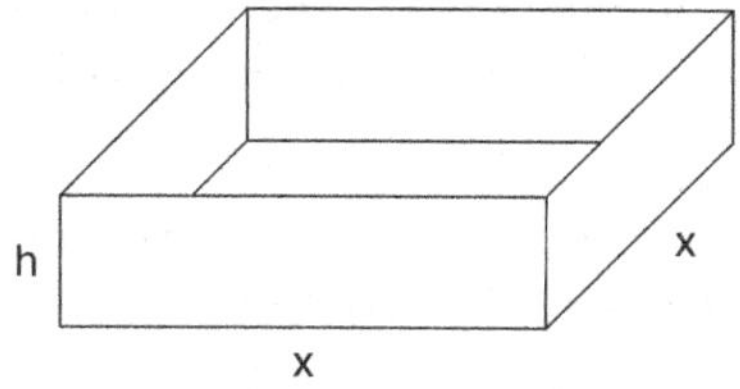

Letting x = length of the square base and h = height of the box, you find that

$$V = x^2 h = 108 \text{ cu in} \Rightarrow h = \frac{108}{x^2}$$

$$S = x^2 + 4xh$$

$$S = f(x) = x^2 + 4x\left(\frac{108}{x^2}\right)$$

$$f(x) = x^2 + \frac{432}{x}$$

with the domain of $f(x) = (0,+\infty)$ because x represents a length.

$$f'(x) = 2x - \frac{432}{x^2}$$

$$f'(x) = 0 \Rightarrow 2x - \frac{432}{x^2} = 0$$

$$2x^3 - 432 = 0$$

$$2x^3 = 432$$

$$x^3 = 216$$

$$x = 6$$

hence, a critical point occurs when $x = 6$. Using the Second Derivative Test:

$$f''(x) = 2 + \frac{864}{x^3}$$

$$f''(6) = 6 > 0$$

and f has a relative minimum at $x = 6$; hence, the dimensions of the box that require the least amount of material are a length and width of 6 inches and a height of 3 inches.

Example 4-17: A right circular cylinder is inscribed in a right circular cone so that the center lines of the cylinder and the cone coincide. The cone has 8 cm and radius 6 cm. Find the maximum volume possible for the inscribed cylinder.

The function that is to be maximized is the volume (V) of a cylinder inscribed in a cone with height 8 cm and radius 6 cm (Figure 4-3).

Letting r = radius of the cylinder and h = height of the cylinder and applying similar triangles, you find that

$$\frac{h}{8} = \frac{6-r}{6}$$

$$6h = 48 - 8r$$

$$h = 8 - \frac{4}{3}r$$

Figure 4-3 A cross section of the cone and cylinder for Example 4-17.

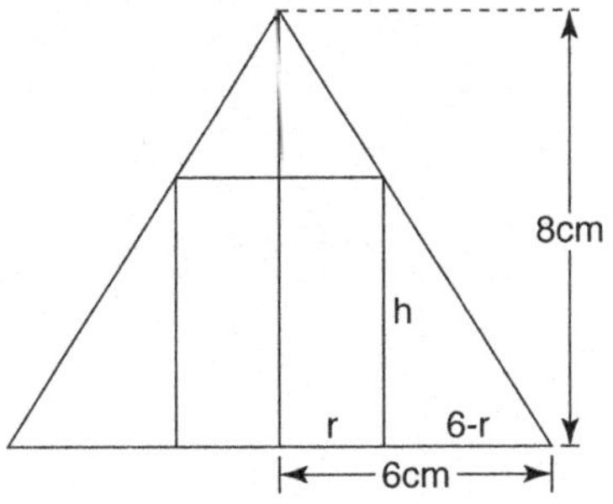

Because $V = \pi r^2 h$ and $h = 8 - (4/3)r$, you find that

$$V = f(r) = \pi r^2 (8 - \frac{4}{3} r)$$

$$f(r) = 8\pi r^2 - \frac{4}{3}\pi r^3$$

with the domain of $f(r) = [0,6]$ because r represents the radius of the cylinder, which cannot be greater that the radius of the cone.

$$f'(r) = 16\pi r - 4\pi r^2$$

$$f'(r) = 0 \Rightarrow 16\pi r - 4\pi r^2 = 0$$

$$4\pi r(4 - r) = 0$$

$$r = 0, 4$$

Because $f(r)$ is continuous on [0,6], use the Extreme Value Theorem and evaluate the function at its critical points and its endpoints; hence,

$$f(0) = 0$$

$$f(4) = \frac{128\pi}{3}$$

$$f(6) = 0$$

hence, the maximum volume is $128\pi/3\ \text{cm}^3$, which will occur when the radius of the cylinder is 4 cm and its height is 8/3 cm.

Distance, Velocity, and Acceleration

As previously mentioned, the derivative of a function representing the position of a particle along a line at time t is the instantaneous velocity at that time. The derivative of the velocity, which is the second derivative of the position function, represents the *instantaneous acceleration* of the particle at time t.

If $y = s(t)$ represents the position function, then $v = s'(t)$ represents the instantaneous velocity, and $a = v'(t) = s''(t)$ represents the instantaneous acceleration of the particle at time t.

A positive velocity indicates that the position is increasing as time increases, while a negative velocity indicates that the position is decreasing with respect to time. If the distance remains constant, then the velocity will be zero on such an interval of time. Likewise, a positive acceleration implies that the velocity is increasing with respect to time, and a negative acceleration implies that the velocity is decreasing with respect to time. If the velocity remains constant on an interval of time, then the acceleration will be zero on the interval.

Example 4-18: The position of a particle on a line is given by $s(t) = t^3 - 3t^2 - 6t + 5$, where t is measured in seconds and s is measured in feet. Find

(a) The velocity of the particle at the end of 2 seconds.

(b) The acceleration of the particle at the end of 2 seconds.

Part (a): The velocity of the particle is

$$v = s'(t) = 3t^2 - 6t - 6$$

$$\text{At } t = 2 \text{ seconds } s'(2) = 3(2)^2 - 6(2) - 6$$

$$s'(2) = -6 \text{ ft/sec}$$

Part (b): The acceleration of the particle is

$$a = v'(t) = s''(t) = 6t - 6$$

$$\text{At } t = 2 \text{ seconds } v'(2) = s''(2) = 6(2) - 6$$

$$v'(2) = s''(2) = 6 \text{ ft/sec}^2$$

Example 4-19: The formula $s(t) = -4.9t^2 + 49t + 15$ gives the height in meters of an object after it is thrown vertically upward from a point 15 meters above the ground at a velocity of 49 m/sec. How high above the ground will the object reach?

The velocity of the object will be zero at its highest point above the ground. That is, $v = s'(t) = 0$, where

$$v = s'(t) = -9.8t + 49$$

$$s'(t) = 0 \Rightarrow -9.8t + 49 = 0$$

$$-9.8t = -49$$

$$t = 5 \text{ seconds}$$

The height above the ground at 5 seconds is

$$s(5) = -4.9(5)^2 + 49(5) + 15$$

$$s(5) = 137.5 \text{ meters}$$

hence, the object will reach its highest point at 137.5 m above the ground.

Related Rates of Change

Some problems in calculus require finding the rate of change or two or more variables that are related to a common variable, namely time. To solve these types of problems, the appropriate rate of change is determined by implicit differentiation with respect to time. Note that a given rate of change is positive if the dependent variable increases with respect to time and negative if the dependent variable decreases with respect to time. The sign of the rate of change of the solution variable with respect to time will also indicate whether the variable is increasing or decreasing with respect to time.

Example 4-20: Air is being pumped into a spherical balloon such that its radius increases at a rate of .75 in/min. Find the rate of change of its volume when the radius is 5 inches.

The volume (V) of a sphere with radius r is

$$V = \frac{4}{3}\pi r^3$$

Differentiating with respect to t, you find that

$$\frac{dV}{dt} = \frac{4}{3}\pi \cdot 3r^2 \cdot \frac{dr}{dt}$$
$$\frac{dV}{dt} = 4\pi r^2 \cdot \frac{dr}{dt}$$

The rate of change of the radius dr/dt = .75 in/min because the radius is increasing with respect to time.

At $r = 5$ inches, you find that

$$\frac{dV}{dt} = 4\pi\,(5\,\text{inches})^2 \cdot (.75\,\text{in/min})$$
$$\frac{dV}{dt} = 75\pi\ \text{in}^3/\text{min}$$

hence, the volume is increasing at a rate of 75π in^3/min when the radius has a length of 5 inches.

Example 4-21: A car is traveling north toward an intersection at a rate of 60 mph while a truck is traveling east away from the intersection at a rate of 50 mph. Find the rate of change of the distance between the car and truck when the car is 3 miles south of the intersection and the truck is 4 miles east of the intersection.

Let x = distance traveled by the truck
y = distance traveled by the car
z = distance between the car and truck

The distances are related by the Pythagorean Theorem: $x^2 + y^2 = z^2$ (Figure 4-4).

Figure 4-4 A diagram of the situation for Example 4-21.

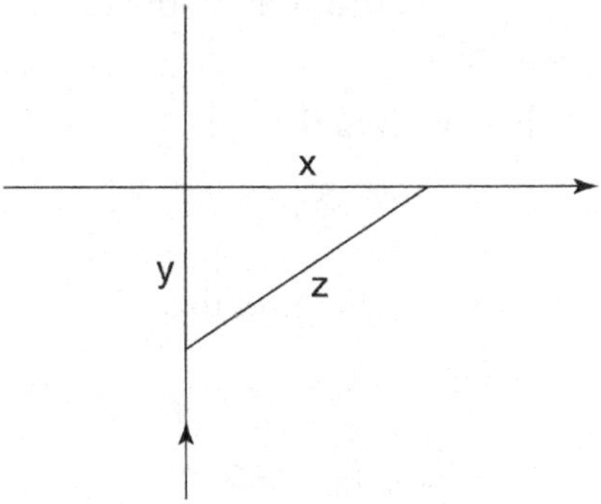

The rate of change of the truck is $dx/dt = 50$ mph because it is traveling away from the intersection, while the rate of change of the car is $dy/dt = -60$ mph because it is traveling toward the intersection. Differentiating with respect to time, you find that

$$2x\frac{dy}{dt} + 2y\frac{dy}{dt} = 2z\frac{dz}{dt}$$

$$x\frac{dx}{dt} + y\frac{dy}{dt} = z\frac{dz}{dt}$$

$$(4\text{ mi})(50\text{ mph}) + (3\text{ mi})(-60\text{ mph}) = (5\text{ mi})\frac{dz}{dt}$$

$$20\text{ mi}^2\text{ph} = (5\text{ mi})\frac{dz}{dt}$$

$$\frac{dz}{dt} = 4\text{ mph}$$

hence, the distance between the car and the truck is increasing at a rate of 4 mph at the time in question.

Differentials

The derivative of a function can often be used to approximate certain function values with a surprising degree of accuracy. To do this, the concept of the differential of the independent variable and the dependent variable must be introduced.

The definition of the derivative of a function $y = f(x)$ as you recall is

$$f'(x) = \lim_{w \to x} \frac{f(w) - f(x)}{w - x} = \lim_{\Delta x \to 0} \frac{\Delta y}{\Delta x}$$

where $\Delta y = f(w) - f(x)$ and $\Delta x = w - x$, so $f(w) = f(x + \Delta x)$.

The derivative $f'(x)$ represents the slope of the tangent line to the curve at some point $(x, f(x))$. If Δx is very small ($\Delta x \neq 0$), then the slope of the tangent is approximately the same as the slope of the secant line through $(x, f(x))$. That is,

$$f'(x) \approx \left[f(x+\Delta x) - f(x)\right]/\Delta x$$

$$\text{or equivalently } f'(x) \cdot \Delta x \approx f(x+\Delta x) - f(x)$$

The differential of the independent variable x is written dx and is the same as the change in x, Δx. That is,

$$dx = \Delta x, \Delta x \neq 0$$

$$\text{hence, } f'(x) \cdot dx \approx f(x+\Delta x) - f(x)$$

The differential of the dependent variable y, written dy, is defined to be

$$dy = f'(x) \cdot dx \approx f(x+\Delta x) - f(x)$$

$$\text{Because } \Delta y = f(x+\Delta x) - f(x)$$

$$\text{you find that } dy = f'(x)\, dx \approx \Delta y$$

The conclusion to be drawn from the preceding discussion is that the differential of $y(dy)$ is approximately equal to the exact change in $y(\Delta y)$, provided that the change in x $(\Delta x = dx)$ is relatively small. The smaller the change in x, the closer dy will be to Δy, enabling you to approximate function values close to $f(x)$ (Figure 4-5).

Figure 4-5 Approximating a function with differentials.

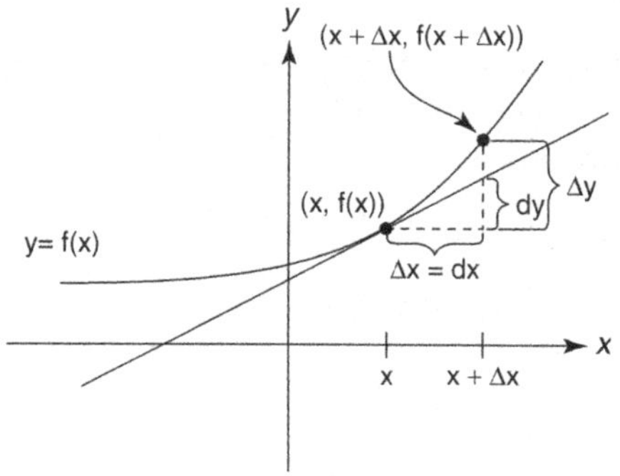

Example 4-22: Find dy for $y = x^3 + 5x - 1$.

$$\text{Because } y = f(x) = x^3 + 5x - 1$$
$$f'(x) = 3x^2 + 5$$
$$dy = f'(x) \cdot dx$$
$$dy = (3x^2 + 5) \cdot dx$$

Example 4-23: Use differentials to approximate the change in the area of a square if the length of its side increases from 6 cm to 6.23 cm.

Let x = length of the side of the square. The area may be expressed as a function of x, where $y = f(x) = x^2$. The differential dy is

$$dy = f'(x) \cdot dx$$
$$dy = 2x \cdot dx$$

Because x is increasing from 6 to 6.23, you find that $\Delta x = dx = .23$ cm; hence,

$$dy = 2\,(6\text{ cm})(.23\text{ cm})$$
$$dy = 2.76\text{ cm}^2$$

The area of the square will increase by approximately 2.76 cm^2 as its side length increases from 6 to 6.23. Note that the exact increase in area (Δy) is 2.8129 cm^2.

Example 4-24: Use differentials to approximate the value of $\sqrt[3]{26.55}$ to the nearest thousandth.

Because the function you are applying is $f(x) = \sqrt[3]{x}$, choose a convenient value of x that is a perfect cube and is relatively close to 26.55, namely $x = 27$. The differential dy is

$$dy = f'(x)\,dx$$
$$dy = \frac{1}{3}x^{-2/3}\,dx$$
$$dy = \frac{1}{3x^{2/3}}\,dx$$

Because x is decreasing from 27 to 26.55, you find that $\Delta x = dx = -.45$; hence,

$$dy = \frac{1}{3(27)^{2/3}} \cdot (-.45)$$

$$= \frac{1}{27} \cdot -\frac{45}{100}$$

$$dy = -\frac{1}{60}$$

which implies that $\sqrt[3]{26.55}$ will be approximately 1/60 less that $\sqrt[3]{27} = 3$; hence,

$$\sqrt[3]{26.55} \approx 3 - \frac{1}{60}$$

$$\approx 3 - .0167$$

$$\approx 2.9833$$

$$\sqrt[3]{26.55} \approx 2.983 \text{ to the nearest thousandth}$$

Note that the calculator value of $\sqrt[3]{26.55}$ is 2.983239874, which rounds to the same answer to the nearest thousandth!

Chapter Checkout

Q&A

1. Find equations for the tangent and normal lines to $y = \frac{x}{x^2 + 1}$ at the point $(2, \frac{2}{5})$.

2. For the function $y = x^3 - 5x^2 + 5$ on [0,6], find

(a) The maximum and minimum values of the function.

(b) All intervals where the function is increasing or decreasing.

(c) The concavity and any inflection points of the function.

3. A right circular cylinder is to be made with a volume of 100π cubic inches. Find the dimensions for the cylinder that require the least amount of material.

4. The formula $s(t) = -4.9t^2 + 20t + 2$ gives the height in meters of an object after it is thrown vertically upward from a point 2 meters above the ground at a velocity of 20 m/sec. How high above the ground will the object reach?

5. Air is being pumped into a spherical balloon such that its volume increases at a rate of 2 in^3/sec. Find the rate of change of its radius when the radius is 6 inches.

Answers: 1. tangent line $3x + 25y = 16$, normal line $25x - 3y = 244/5$ **2.** (a) maximum 41, minimum –365/27 (b) increasing on (10/3, 6], decreasing on (0,10/3) (c) concave down on [0,5/3), concave up on (5/3, 6], inflection point at $(5/3, -115/27)$ **3.** $r = \sqrt[3]{50}$, $h = 2 \cdot \sqrt[3]{50}$ **4.** approximately 22.4 meters **5.** $1/(72\pi)$ in/sec

Chapter 5

INTEGRATION

Chapter Check-In

- ❑ Understanding and computing basic indefinite integrals
- ❑ Using more advanced techniques of integration
- ❑ Understanding and computing definite integrals

Along with differentiation, a second important operation of calculus is *antidifferentiation,* or *integration.* These operations may be thought of as inverse of one another, and the rules for finding derivatives discussed in previous chapters will be useful in establishing corresponding rules for finding antiderivatives. The relationship between antiderivatives and definite integrals is discussed later in the chapter with the statement of the Fundamental Theorem of Calculus.

Antiderivatives/Indefinite Integrals

A function $F(x)$ is called an **antiderivative** of a function of $f(x)$ if $F'(x) = f(x)$ for all x in the domain of f. Note that the function F is not unique and that an infinite number of antiderivatives could exist for a given function. For example, $F(x) = x^3$, $G(x) = x^3 + 5$, and $H(x) = x^3 - 2$ are all antiderivatives of $f(x) = 3x^2$ because $F'(x) = G'(x) = H'(x) = f(x)$ for all x in the domain of f. It is clear that these functions F, G, and H differ only by some constant value and that the derivative of that constant value is always zero. In other words, if $F(x)$ and $G(x)$ are antiderivatives of $f(x)$ on some interval, then $F'(x)= G'(x)$ and $F(x) = G(x) + C$ for some constant C in the interval. Geometrically, this means that the graphs of $F(x)$ and $G(x)$ are identical except for their vertical position.

The notation used to represent all antiderivatives of a function $f(x)$ is the **indefinite integral** symbol written ($\int$), where $\int f(x)\,dx = F(x) + C$. The function of $f(x)$ is called the integrand, and C is referred to as the constant of integration. The expression $F(x) + C$ is called the indefinite integral of

F with respect to the independent variable x. Using the previous example of $F(x) = x^3$ and $f(x) = 3x^2$, you find that $\int 3x^2\,dx = x^3 + C$.

The indefinite integral of a function is sometimes called the general antiderivative of the function as well.

Example 5-1: Find the indefinite integral of $f(x) = \cos x$.

Because the derivative of $F(x) = \sin x$ is $F'(x) = \cos x$, write

$$\int \cos x\,dx = \sin x + C.$$

Example 5-2: Find the general antiderivative of $f(x) = -8$.

Because the derivative of $F(x) = -8x$ is $F'(x) = -8$, write

$$\int -8dx = -8x + C.$$

Integration Techniques

Many integration formulas can be derived directly from their corresponding derivative formulas, while other integration problems require more work. Some that require more work are substitution and change of variables, integration by parts, trigonometric integrals, and trigonometric substitutions.

Basic formulas

Most of the following basic formulas directly follow the differentiation rules that were discussed in preceding chapters.

1. $\int kf(x)\,dx = k\int f(x)\,dx$
2. $\int [f(x) \pm g(x)]dx = \int f(x)\,dx \pm \int g(x)\,dx$
3. $\int kdx = kx + C$
4. $\int x^n\,dx = \dfrac{x^{n+1}}{n+1} + C, n \neq -1$
5. $\int \sin x\,dx = -\cos x + C$
6. $\int \cos x\,dx = \sin x + C$
7. $\int \sec^2 x\,dx = \tan x + C$

8. $\int \csc^2 x\,dx = -\cot x + C$

9. $\int \sec x \tan x\,dx = \sec x + C$

10. $\int \csc x \cot x\,dx = -\csc x + C$

11. $\int e^x\,dx = e^x + C$

12. $\int a^x\,dx = \frac{a^x}{\ln a} + C, a > 0, a \neq 1$

13. $\int \frac{dx}{x} = \ln|x| + C$

14. $\int \tan x\,dx = \ln|\sec x| + C$

15. $\int \cot x\,dx = \ln|\sin x| + C$

16. $\int \sec x\,dx = \ln|\sec x + \tan x| + C$

17. $\int \csc x\,dx = -\ln|\csc x + \cot x| + C$

18. $\int \frac{dx}{\sqrt{a^2 - x^2}} = \sin^{-1}\frac{x}{a} + C$

19. $\int \frac{dx}{a^2 + x^2} = \frac{1}{a}\tan^{-1}\frac{x}{a} + C$

20. $\int \frac{dx}{x\sqrt{x^2 - a^2}} = \frac{1}{a}\sec^{-1}\left|\frac{x}{a}\right| + C$

Example 5-3: Evaluate $\int x^4\,dx$.

Using formula (4) from the preceding list, you find that $\int x^4\,dx = \frac{x^5}{5} + C$.

Example 5-4: Evaluate $\int \frac{1}{\sqrt{x}}\,dx$.

Because $1/\sqrt{x} = x^{-1/2}$, using formula (4) from the preceding list yields

$$\int \frac{1}{\sqrt{x}}\,dx = \int x^{-1/2}\,dx$$

$$= \frac{x^{1/2}}{\frac{1}{2}} + C$$

$$= 2x^{1/2} + C$$

Example 5-5: Evaluate $\int (6x^2 + 5x - 3)\, dx$

Applying formulas (1), (2), (3), and (4), you find that

$$\int (6x^2 + 5x - 3)\, dx = \frac{6x^3}{3} + \frac{5x^2}{2} - 3x + C$$
$$= 2x^3 + \frac{5}{2}x^2 - 3x + C$$

Example 5-6: Evaluate $\int \frac{dx}{x+4}$.

Using formula (13), you find that $\int \frac{dx}{x+4} = \ln|x+4| + C$.

Example 5-7: Evaluate $\int \frac{dx}{25 + x^2}$.

Using formula (19) with a = 5, you find that

$$\int \frac{dx}{25 + x^2} = \frac{1}{5}\tan^{-1}\frac{x}{5} + C$$

Substitution and change of variables

One of the integration techniques that is useful in evaluating indefinite integrals that do not seem to fit the basic formulas is **substitution and change of variables.** This technique is often compared to the chain rule for differentiation because they both apply to composite functions. In this method, the inside function of the composition is usually replaced by a single variable (often u). Note that the derivative or a constant multiple of the derivative of the inside function must be a factor of the integrand.

The purpose in using the substitution technique is to rewrite the integration problem in terms of the new variable so that one or more of the basic integration formulas can then be applied. Although this approach may seem like more work initially, it will eventually make the indefinite integral much easier to evaluate.

It is usually desirable to express the final answer in terms of the original variable of integration.

Example 5-8: Evaluate $\int x^2 (x^3 + 1)^5\, dx$.

Because the inside function of the composition is $x^3 + 1$, substitute with

$$u = x^3 + 1$$
$$du = 3x^2\, dx$$
$$\frac{1}{3}du = x^2\, dx$$

hence,
$$\int x^2(x^3+1)^5\,dx = \frac{1}{3}\int u^5\,du$$
$$= \frac{1}{3}\cdot\frac{u^6}{6} + C$$
$$= \frac{1}{18}u^6 + C$$
$$= \frac{1}{18}(x^3+1)^6 + C$$

Example 5-9: Evaluate $\int \sin(5x)\,dx$.

Because the inside function of the composition is $5x$, substitute with
$$u = 5x$$
$$du = 5dx$$
$$\frac{1}{5}du = dx$$

hence,
$$\int \sin(5x)\,dx = \frac{1}{5}\int \sin u\,du$$
$$= -\frac{1}{5}\cos u + C$$
$$= -\frac{1}{5}\cos(5x) + C$$

Example 5-10: Evaluate $\int \frac{3x}{\sqrt{9-x^2}}\,dx$.

Because the inside function of the composition is $9 - x^2$, substitute with
$$u = 9 - x^2$$
$$du = -2x\,dx$$
$$-\frac{1}{2}du = x\,dx$$

hence,
$$\int \frac{3x}{\sqrt{9-x^2}}\,dx = -\frac{3}{2}\int \frac{1}{\sqrt{u}}\,du$$
$$= -\frac{3}{2}\int u^{-1/2}\,du$$
$$= -\frac{3}{2}\cdot\frac{u^{1/2}}{\frac{1}{2}} + C$$
$$= -3u^{1/2} + C$$
$$= -3\sqrt{9-x^2} + C$$

Integration by parts 8.2

Another integration technique to consider in evaluating indefinite integrals that do not fit the basic formulas is **integration by parts.** You may consider this method when the integrand is a single transcendental function or a product of an algebraic function and a transcendental function. The basic formula for integration by parts is

$$\int u\,dv = uv - \int v\,du$$

where u and v are differential functions of the variable of integration.

A general rule of thumb to follow is to first choose dv as the most complicated part of the integrand that can be easily integrated to find v. The u function will be the remaining part of the integrand that will be differentiated to find du. The goal of this technique is to find an integral, $\int v\,du$, which is easier to evaluate than the original integral.

Example 5-11: Evaluate $\int x\sec^2 x\,dx$.

$$\text{Let } u = x \text{ and } dv = \sec^2 x\,dx$$

$$du = dx \quad v = \tan x$$

hence,

$$\int x\sec^2 x\,dx = x\tan x - \int \tan x\,dx$$

$$= x\tan x - \ln|\sec x| + C$$

Example 5-12: Evaluate $\int x^4 \ln x\,dx$.

$$\text{Let } u = \ln x \text{ and } dv = x^4\,dx$$

$$du = \frac{1}{x}\,dx \quad v = \frac{x^5}{5}$$

hence,

$$\int x^4 \ln x\,dx = \frac{x^5}{5}\ln x - \int \frac{x^5}{5}\cdot\frac{1}{x}\,dx$$

$$= \frac{x^5}{5}\ln x - \frac{1}{5}\int x^4\,dx$$

$$= \frac{1}{5}x^5 \ln x - \frac{1}{25}x^5 + C$$

Example 5-13: Evaluate $\int \tan^{-1} x\,dx$.

$$\text{Let } u = \tan^{-1} x \text{ and } dv = dx$$

$$du = \frac{1}{1+x^2}\,dx \quad v = x$$

hence,

$$\int \tan^{-1} x\,dx = x\tan^{-1} x - \int \frac{x}{1+x^2}\,dx$$

$$= x\tan^{-1} x - \frac{1}{2}\ln(1+x^2) + C$$

Trigonometric integrals

Integrals involving powers of the trigonometric functions must often be manipulated to get them into a form in which the basic integration formulas can be applied. It is extremely important for you to be familiar with the basic trigonometric identities that were reviewed in Chapter 1 because you often used these to rewrite the integrand in a more workable form. As in integration by parts, the goal is to find an integral that is easier to evaluate than the original integral.

Example 5-14: Evaluate $\int \cos^3 x \sin^4 dx$

$$\int \cos^3 \sin^4 x\,dx = \int \cos^2 x \sin^4 x \cos x\,dx$$

$$= \int (1 - \sin^2 x)\sin^4 x \cos x\,dx$$

$$= \int (\sin^4 x - \sin^6 x)\cos x\,dx$$

$$= \int \sin^4 x \cos x\,dx - \int \sin^6 x \cos x\,dx$$

$$= \frac{1}{5}\sin^5 x - \frac{1}{7}\sin^7 x + C$$

Example 5-15: Evaluate $\int \sec^6 x\,dx$

$$\begin{aligned}\int \sec^6 x\,dx &= \int \sec^4 x \sec^2 x\,dx\\ &= \int (\sec^2 x)^2 \sec^2 x\,dx\\ &= \int (\tan^2 x + 1)^2 \sec^2 x\,dx\\ &= \int (\tan^4 x + 2\tan^2 x + 1)\sec^2 x\,dx\\ &= \int \tan^4 x \sec^2 x\,dx + \int 2\tan^2 x \sec^2 x\,dx + \int \sec^2 x\,dx\\ &= \frac{1}{5}\tan^5 x + \frac{2}{3}\tan^3 x + \tan x + C\end{aligned}$$

Example 5-16: Evaluate $\int \sin^4 x\,dx.$

$$\begin{aligned}\int \sin^4 x\,dx &= \int (\sin^2 x)^2\,dx\\ &= \int \left(\frac{1-\cos 2x}{2}\right)^2 dx\\ &= \frac{1}{4}\int (1 - 2\cos 2x + \cos^2 2x)\,dx\\ &= \frac{1}{4}\int \left(1 - 2\cos 2x + \frac{1+\cos 4x}{2}\right) dx\\ &= \frac{1}{4}\int \left(\frac{3}{2} - 2\cos 2x + \frac{\cos 4x}{2}\right) dx\\ &= \frac{1}{8}\int (3 - 4\cos 2x + \cos 4x)\,dx\\ &= \frac{1}{8}\left(3x - 2\sin 2x + \frac{1}{4}\sin 4x\right) + C\\ &= \frac{3}{8}x - \frac{1}{4}\sin 2x + \frac{1}{32}\sin 4x + C\end{aligned}$$

Trigonometric substitutions

If an integrand contains a radical expression of the form $\sqrt{a^2-x^2}, \sqrt{a^2+x^2}, \text{or} \sqrt{x^2-a^2}$, a specific **trigonometric substitution** may be helpful in evaluating the indefinite integral. Some general rules to follow are

1. If the integrand contains $\sqrt{a^2-x^2}$
   let $x = a \sin \theta$
   $dx = a \cos \theta d\theta$
   and $\sqrt{a^2-x^2} = a\cos\theta$
2. If the integrand contains $\sqrt{a^2+x^2}$
   let $x = a \tan \theta$
   $dx = a \sec^2 \theta d\theta$
   and $\sqrt{a^2+x^2} = a\sec\theta$
3. If the integrand contains $\sqrt{x^2-a^2}$
   let $x = a \sec \theta$
   $dx = a \sec \theta \tan\theta d\theta$
   and $\sqrt{x^2-a^2} = a\tan\theta$

Right triangles may be used in each of the three preceding cases to determine the expression for any of the six trigonometric functions that appear in the evaluation of the indefinite integral.

Example 5-17: Evaluate $\int \frac{dx}{x^2\sqrt{4-x^2}}$.

Because the radical has the form $\sqrt{a^2-x^2}$

$$\text{let } x = a\sin\theta = 2\sin\theta$$

$$dx = 2\cos\theta d\theta$$

$$\text{and } \sqrt{4-x^2} = 2\cos\theta \text{ (Figure 5 - 1)}$$

Figure 5-1 Diagram for Example 5-17.

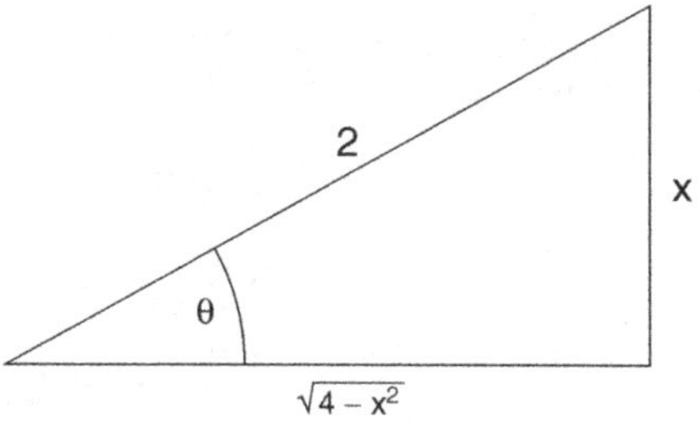

hence,
$$\int \frac{dx}{x^2\sqrt{4-x^2}} = \int \frac{2\cos\theta d\theta}{(4\sin^2\theta)(2\cos\theta)}$$
$$= \frac{1}{4}\int \frac{d\theta}{\sin^2\theta}$$
$$= \frac{1}{4}\int \csc^2\theta d\theta$$
$$= -\frac{1}{4}\cot\theta + C$$
$$= -\frac{1}{4}\cdot\frac{\sqrt{4-x^2}}{x} + C$$
$$= -\frac{\sqrt{4-x^2}}{4x} + C$$

Example 5-18: Evaluate $\int \frac{dx}{\sqrt{25+x^2}}$.

Because the radical has the form $\sqrt{a^2+x^2}$

let $x = a\tan\theta = 5\tan\theta$

$dx = 5\sec^2\theta d\theta$

and $\sqrt{25+x^2} = 5\sec\theta$ (Figure 5 − 2)

Figure 5-2 Diagram for Example 5-18.

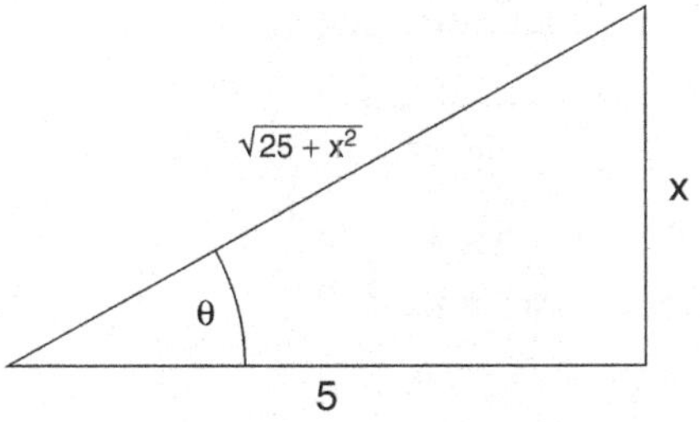

hence,
$$\int \frac{dx}{\sqrt{25+x^2}} = \int \frac{5\sec^2\theta d\theta}{5\sec\theta}$$
$$= \int \sec\theta d\theta$$
$$= \ln|\sec\theta + \tan\theta| + C$$
$$= \ln\left|\frac{\sqrt{25+x^2}}{5} + \frac{x}{5}\right| + C$$

Distance, Velocity, and Acceleration

The indefinite integral is commonly applied in problems involving distance, velocity, and acceleration, each of which is a function of time. In the discussion of the applications of the derivative, note that the derivative of a distance function represents *instantaneous velocity* and that the derivative of the velocity function represents *instantaneous acceleration* at a particular time. In considering the relationship between the derivative and the indefinite integral as inverse operations, note that the indefinite integral of the acceleration function represents the velocity function and that the indefinite integral of the velocity represents the distance function.

In case of a free-falling object, the acceleration due to gravity is –32 ft/sec^2. The significance of the negative is that the rate of change of the velocity with respect to time (acceleration), is negative because the velocity is decreasing as the time increases. Using the fact that the velocity is the indefinite integral of the acceleration, you find that

$$\begin{aligned} a(t) &= s''(t) = -32 \\ v(t) &= s'(t) = \int s''(t)\,dt \\ &= \int -32dt \\ &= -32t + C_1 \end{aligned}$$

Now, at $t = 0$, the initial velocity (v_0) is

$$\begin{aligned} v_0 &= v(0) = (-32)(0) + C_1 \\ v_0 &= C_1 \end{aligned}$$

hence, because the constant of integration for the velocity in this situation is equal to the initial velocity, write $v(t) = -32t + v_0$.

Because the distance is the indefinite integral of the velocity, you find that

$$\begin{aligned} s(t) &= \int v(t)\,dt \\ &= \int (-32t + v_0)\,dt \\ &= -32 \cdot \frac{t^2}{2} + v_0 t + C_2 \\ &= -16t^2 + v_0 t + C_2 \end{aligned}$$

Now, at $t = 0$, the initial distance (s_0) is

$$\begin{aligned} s_0 &= s(0) = -16(0)^2 + v_0(0) + C_2 \\ s_0 &= C_2 \end{aligned}$$

hence, because the constant of integration for the distance in this situation is equal to the initial distance, write $s(t) = -16t^2 + v_0\,(t) + s_0$.

Example 5-19: A ball is thrown downward from a height of 512 feet with a velocity of 64 feet per second. How long will it take for the ball to reach the ground?

From the given conditions, you find that

$$a(t) = -32 \text{ ft/sec}^2$$
$$v_0 = -64 \text{ ft/sec}$$
$$s_0 = 512 \text{ ft}$$

hence,

$$v(t) = -32t - 64$$
$$s(t) = -16t^2 - 64t + 512$$

The distance is zero when the ball reaches the ground or

$$-16t^2 - 64t + 512 = 0$$
$$-16(t^2 + 4t - 32) = 0$$
$$-16(t + 8)(t - 4) = 0$$
$$t = -8,\ t = 4$$

hence, the ball will reach the ground 4 seconds after it is thrown.

Example 5-20: In the previous example, what will the velocity of the ball be when it hits the ground?

Because $v(t) = -32(t) - 64$ and it takes 4 seconds for the ball to reach the ground, you find that

$$v(4) = -32(4) - 64$$
$$= -192 \text{ ft/sec}$$

hence, the ball will hit the ground with a velocity of –192 ft/sec. The significance of the negative velocity is that the rate of change of the distance with respect to time (velocity) is negative because the distance is decreasing as the time increases.

Example 5-21: A missile is accelerating at a rate of $4t$ m/sec^2 from a position at rest in a silo 35 m below ground level. How high above the ground will it be after 6 seconds?

From the given conditions, you find that $a(t) = 4t$ m/sec^2, $v_0 = 0$ m/sec because it begins at rest, and $s_0 = -35$ m because the missile is below ground level; hence,

$$v(t) = \int 4t\,dt = 2t^2$$

and $$s(t) = \int 2t^2\,dt = \frac{2}{3}t^3 - 35$$

After 6 seconds, you find that $s(6) = \frac{2}{3}(6)^3 - 35\text{m} = 109\text{m}$

hence, the missile will be 109 m above the ground after 6 seconds.

Definite Integrals

The **definite integral** of a function is closely related to the antiderivative and indefinite integral of a function. The primary difference is that the definite integral, if it exists, is a real number value, while the latter two represent an infinite number of functions that differ only by a constant. The relationship between these concepts is will be discussed in the section on the Fundamental Theorem of Calculus, and you will see that the definite integral will have applications to many problems in calculus.

Definition of definite integrals

The development of the definition of the definite integral begins with a function $f(x)$, which is continuous on a closed interval $[a,b]$. The given interval is partitioned into "n" subintervals that, although not necessary, can be taken to be of equal lengths (Δx). An arbitrary domain value, x_i^*, is chosen in each subinterval, and its subsequent function value, $f(x_i)$, is determined. The product of each function value times the corresponding subinterval length is determined, and these "n" products are added to determine their sum. This sum is referred to as a **Riemann sum** and may be positive, negative, or zero, depending upon the behavior of the function on the closed interval. For example, if $f(x) > 0$ on $[a,b]$, then the Riemann sum will be a positive real number. If $f(x) < 0$ on $[a,b]$, then the Riemann sum will be a negative real number. The Riemann sum of the function $f(x)$ on $[a,b]$ is expressed as

$$S_n = f(x_1^*)\Delta x + f(x_2^*)\Delta x + f(x_3^*)\Delta x + \cdots + f(x_n^*)\Delta x$$

$$\text{or } S_n = \sum_{i=1}^{n} f(x_i^*)\Delta x$$

A Riemann sum may, therefore, be thought of as a "sum of n products."

Example 5-22: Evaluate the Riemann sum for $f(x) = x^2$ on $[1,3]$ using the four subintervals of equal length, where x_i^* is the right endpoint in the ith subinterval (see Figure 5-3).

Because the subintervals are to be of equal lengths, you find that

$$\begin{aligned}\Delta x &= \frac{b-a}{n}\\ &= \frac{3-1}{4}\\ &= \frac{1}{2}\end{aligned}$$

The Riemann sum for four subintervals is

$$\begin{aligned}S_4 &= \sum_{i=1}^{4} f(x_i^*)\,\Delta x\\ &= f(x_1^*)\,\Delta x + f(x_2^*)\,\Delta x + f(x_3^*)\,\Delta x + f(x_4^*)\,\Delta x\\ &= [f(x_1^*) + f(x_2^*) + f(x_3^*) + f(x_4^*)]\,\Delta x\\ &= \left[f\left(\frac{3}{2}\right) + f(2) + f\left(\frac{5}{2}\right) + f(3)\right]\cdot\frac{1}{2}\\ &= \left[\frac{9}{4} + 4 + \frac{25}{4} + 9\right]\cdot\frac{1}{2}\\ &= \left[\frac{86}{4}\right]\cdot\frac{1}{2}\\ S_4 &= \frac{43}{4}\end{aligned}$$

Figure 5-3 A Riemann sum with four subintervals.

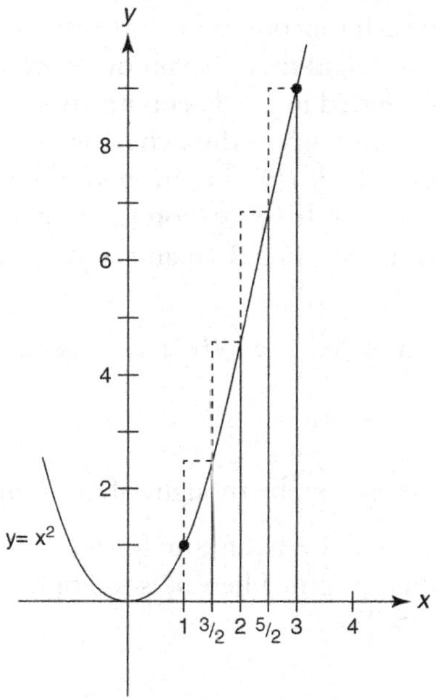

If the number of subintervals is increased repeatedly, the effect would be that the length of each subinterval would get smaller and smaller. This may be restated as follows: If the number of subintervals increases without bound ($n \to +\infty$), then the length of each subinterval approaches zero ($\Delta x \to 0$). This limit of a Riemann sum, if it exists, is used to define the definite integral of a function on $[a,b]$. If $f(x)$ is defined on the closed interval $[a,b]$ then the **definite integral** of $f(x)$ from a to b is defined as

$$\int_a^b f(x)\,dx = \lim_{n \to +\infty} S_n$$

$$= \lim_{n \to +\infty} \sum_{i=1}^{n} f(x_i^*)\,\Delta x$$

$$= \lim_{\Delta x \to 0} \sum_{i=1}^{n} f(x_i^*)\,\Delta x$$

if this limit exits.

The function $f(x)$ is called the integrand, and the variable x is the variable of integration. The numbers a and b are called the limits of integration with a referred to as the lower limit of integration while b is referred to as the upper limit of integration.

Note that the symbol $\int$, used with the indefinite integral, is the same symbol used previously for the indefinite integral of a function. The reason for this will be made more apparent in the following discussion of the Fundamental Theorem of Calculus. Also, keep in mind that the definite integral is a unique real number and does not represent an infinite number of functions that result from the indefinite integral of a function.

The question of the existence of the limit of a Riemann sum is important to consider because it determines whether the definite integral exists for a function on a closed interval. As with differentiation, a significant relationship exists between continuity and integration and is summarized as follows: If a function $f(x)$ is continuous on a closed interval $[a,b]$, then the definite integral of $f(x)$ on $[a,b]$ exists and f is said to be integrable on $[a,b]$. In other words, continuity guarantees that the definite integral exists, but the converse is not necessarily true.

Unfortunately, the fact that the definite integral of a function exists on a closed interval does not imply that the value of the definite integral is easy to find.

Properties of definite integrals 5.5,5.6/ET6.5,6.6

Certain properties are useful in solving problems requiring the application of the definite integral. Some of the more common properties are

1. $\int_a^a f(x)\,dx = 0$
2. $\int_a^b f(x)\,dx = -\int_b^a f(x)\,dx$
3. $\int_a^b c\,dx = c(b-a)$, where c is a constant
4. $\int_a^b cf(x)\,dx = c\int_a^b f(x)\,dx$
5. Sum Rule: $\int_a^b [f(x)+g(x)]\,dx = \int_a^b f(x)\,dx + \int_a^b g(x)\,dx$
6. Difference Rule: $\int_a^b [f(x)-g(x)]\,dx = \int_a^b f(x)\,dx - \int_a^b g(x)\,dx$
7. If $f(x) \geq 0$ on $[a, b]$, then $\int_a^b f(x)\,dx \geq 0$
8. If $f(x) \leq 0$ on $[a, b]$, then $\int_a^b f(x)\,dx \leq 0$
9. If $f(x) \geq g(x)$ on $[a, b]$, then $\int_a^b f(x)\,dx \geq \int_a^b g(x)\,dx$
10. If a, b, and c are any three points on a closed interval, then
    $\int_a^b f(x)\,dx = \int_a^c f(x)\,dx + \int_c^b f(x)\,dx$
11. The Mean Value Theorem for Definite Integrals: If $f(x)$ is continuous on the closed interval $[a,b]$, then at least one number c exists in the open interval (a, b) such that

    $$\int_a^b f(x)\,dx = f(c)(b-a)$$

    The value of $f(c)$ is called the average or mean value of the function $f(x)$ on the interval $[a, b]$ and

    $$f(x) = \frac{1}{b-a}\int_a^b f(x)\,dx$$

Example 5-23: Evaluate

$$\int_2^6 3\,dx$$

$$\int_2^6 3\,dx = 3(6-2)$$
$$= 12$$

Example 5-24: Given that $\int_0^3 x^2\,dx = 9$, evaluate $\int_0^3 -4x^2\,dx$

$$\begin{aligned}\int_0^3 -4x^2\,dx &= -4\int_0^3 x^2\,dx\\ &= (-4)\cdot 9\\ &= -36\end{aligned}$$

Example 5-25: Given that $\int_4^9 \sqrt{x}\,dx = \frac{38}{3}$, evaluate $\int_9^4 \sqrt{x}\,dx$

$$\begin{aligned}\int_9^4 \sqrt{x}\,dx &= -\int_4^9 \sqrt{x}\,dx\\ &= -\frac{38}{3}\end{aligned}$$

Example 5-26: Evaluate $\int_3^3 (x^3 + 5x^2 - 3x + 11)\,dx$.

$$\int_3^3 (x^3 + 5x^2 - 3x + 11)\,dx = 0$$

Example 5-27: Given that $\int_1^3 f(x)\,dx = 6$ and $\int_1^3 g(x)\,dx = 10$, evaluate $\int_1^3 [f(x) + g(x)]\,dx$

$$\begin{aligned}\int_1^3 [f(x) + g(x)]\,dx &= \int_1^3 f(x)\,dx + \int_1^3 g(x)\,dx\\ &= 6 + 10\\ &= 16\end{aligned}$$

Example 5-28: Given that $\int_3^7 f(x)\,dx = -2$ and $\int_3^7 g(x)\,dx = 9$, evaluate $\int_3^7 [f(x) - g(x)]\,dx$.

$$\begin{aligned}\int_3^7 [f(x) - g(x)]\,dx &= \int_3^7 f(x)\,dx - \int_3^7 g(x)\,dx\\ &= -2 - 9\\ &= -11\end{aligned}$$

Example 5-29: Given that $\int_2^9 f(x)\,dx = 12$ and $\int_6^9 f(x) = 7$, evaluate $\int_2^6 f(x)\,dx$.

$$\int_2^9 f(x)\,dx = \int_2^6 f(x)\,dx + \int_6^9 f(x)\,dx$$

$$\int_2^6 f(x)\,dx = \int_2^9 f(x)\,dx - \int_6^9 f(x)\,dx$$

$$= 12 - 7$$

$$= 5$$

Example 5-30: Given that $\int_3^6 (x^2 - 2)\,dx = 57$ find all c values that satisfy the Mean Value Theorem for the given function on the closed interval.

By the Mean Value Theorem,

$$\int_a^b f(x)\,dx = f(c)(b - a)$$

for some c in (a, b),

and $$f(c) = \frac{1}{b-a}\int_a^b f(x)\,dx$$

hence, $$f(c) = \frac{1}{b-a}\int_3^6 (x^2 - 2)\,dx$$

$$= \frac{1}{3} \cdot 57$$

$$= 19$$

Because $f(x) = x^2 - 2, f(c) = c^2 - 2$

and $c^2 - 2 = 19$

$$c^2 = 21$$

$$c = \pm\sqrt{21}$$

Because $\sqrt{21} \approx 4.58$ is in the interval (3,6), the conclusion of the Mean Value Theorem is satisfied for this value of c.

The Fundamental Theorem of Calculus

The Fundamental Theorem of Calculus establishes the relationship between indefinite and definite integrals and introduces a technique for evaluating definite integrals without using Riemann sums, which is very important because evaluating the limit of Riemann sum can be extremely time-consuming and difficult. The statement of the theorem is: If $f(x)$ is continuous on the interval $[a,b]$, and $F(x)$ is any antiderivative of $f(x)$ on $[a,b]$, then $\int_a^b f(x)\,dx = F(b) - F(a) = F(x)\big]_a^b$.

In other words, the value of the definite integral of a function on $[a,b]$ is the difference of any antiderivative of the function evaluated at the upper limit of integration minus the same antiderivative evaluated at the lower limit of integration. Because the constants of integration are the same for both parts of this difference, they are ignored in the evaluation of the definite integral because they subtract and yield zero. Keeping this in mind, choose the constant of integration to be zero for all definite integral evaluations after Example 5-31.

Example 5-31: Evaluate $\int_2^5 x^2\,dx$.

Because the general antiderivative of x^2 is $(1/3)x^3 + C$, you find that

$$\begin{aligned}\int_2^5 x^2\,dx &= \left[\frac{1}{3}x^3 + C\right]_2^5 \\ &= \left[\frac{1}{3}(5)^3 + C\right] - \left[\frac{1}{3}(2)^3 + C\right] \\ &= \frac{125}{3} - \frac{8}{3} \\ &= 39\end{aligned}$$

Example 5-32: Evaluate $\int_{\pi/3}^{2\pi} \sin x\,dx$.

Because an antiderivative of $\sin x$ is $-\cos x$, you find that

$$\begin{aligned}\int_{\pi/3}^{2\pi} \sin x\,dx &= -\cos x\Big]_{\pi/3}^{2\pi} \\ &= (-1) - \left(\frac{1}{-2}\right) \\ &= -\frac{1}{2}\end{aligned}$$

Example 5-33: Evaluate $\int_1^4 \sqrt{x}\,dx$.

Because $\sqrt{x} = x^{1/2}$, an antiderivative of $x^{1/2}$ is $\frac{2}{3}x^{3/2}$, and you find that

$$\begin{aligned}\int_1^4 x^{1/2}\,dx &= \frac{2}{3}x^{3/2}\Big]_1^4 \\ &= \frac{2}{3}(4)^{3/2} - \frac{2}{3}(1)^{3/2} \\ &= \frac{16}{3} - \frac{2}{3} \\ &= \frac{14}{3}\end{aligned}$$

Example 5-34: Evaluate $\int_1^3 (x^2 - 4x + 1)\, dx$

Because an antiderivative of $x^2 - 4x + 1$ is $(1/3)x^3 - 2x^2 + x$, you find that

$$\begin{aligned}\int_1^3 (x^2 - 4x + 1)\, dx &= \left[\frac{1}{3}x^3 - 2x^2 + x\right]_1^3 \\ &= \left[\frac{1}{3}(3)^3 - 2(3)^2 + 3\right] - \left[\frac{1}{3}(1)^3 - 2(1)^2 + 1\right] \\ &= (-6) - \left(-\frac{2}{3}\right) \\ &= -\frac{16}{3}\end{aligned}$$

Definite integral evaluation

The numerous techniques that can be used to evaluate indefinite integrals can also be used to evaluate definite integrals. The methods of substitution and change of variables, integration by parts, trigonometric integrals, and trigonometric substitution are illustrated in the following examples.

Example 5-35: Evaluate $\int_1^2 \frac{x\, dx}{(x^2 + 2)^3}$

Using the substitution method with

$$\begin{aligned}u &= x^2 + 2 \\ du &= 2x\, dx \\ \frac{1}{2}\, du &= x\, dx\end{aligned}$$

the limits of integration can be converted from x values to their corresponding u values. When $x = 1$, $u = 3$ and when $x = 2$, $u = 6$, you find that

$$\begin{aligned}\int_1^2 \frac{x\, dx}{(x^2 + 2)^3} &= \frac{1}{2}\int_3^6 \frac{du}{u^3} \\ &= \frac{1}{2}\int_3^6 u^{-3}\, du \\ &= \frac{1}{2}\left[-\frac{1}{2}u^{-2}\right]_3^6 \\ &= -\frac{1}{4}\left[(6)^{-2} - (3)^{-2}\right] \\ &= -\frac{1}{4}\left(\frac{1}{36} - \frac{1}{9}\right) \\ &= \frac{1}{48}\end{aligned}$$

Note that when the substitution method is used to evaluate definite integrals, it is not necessary to go back to the original variable if the limits of integration are converted to the new variable values.

Example 5-36: Evaluate $\int_{\pi}^{3\pi/2} \sqrt{\sin x + 1} \cos x \, dx$.

Using the substitution method with $u = \sin x + 1$, $du = \cos x \, dx$, you find that $u = 1$ when $x = \pi$ and $u = 0$ when $x = 3\pi/2$; hence,

$$\begin{aligned}\int_{\pi}^{3\pi/2} \sqrt{\sin x + 1} \cos x \, dx &= \int_{1}^{0} u^{1/2} \, du \\ &= \frac{2}{3} u^{3/2} \Big]_{1}^{0} \\ &= \frac{2}{3}\left[0^{3/2} - 1^{3/2}\right] \\ &= -\frac{2}{3}\end{aligned}$$

Note that you never had to return to the trigonometric functions in the original integral to evaluate the definite integral.

Example 5-37: Evaluate $\int_{\pi/3}^{\pi/2} x \sin x \, dx$.

Using integration by parts with

$u = x$ and $dv = \sin x \, dx$

$du = dx \quad v = -\cos x$

you find that

$$\begin{aligned}\int x \sin x \, dx &= -x\cos x - \int -\cos x \, dx \\ &= -x \cos x + \sin x + C\end{aligned}$$

hence,

$$\begin{aligned}\int_{\pi/3}^{\pi/2} x \sin x \, dx &= \left[-x \cos x + \sin x\right]_{\pi/3}^{\pi/2} \\ &= \left[\left(-\frac{\pi}{2}\right)\left(\cos\frac{\pi}{2}\right) + \sin\frac{\pi}{2}\right] - \left[\left(-\frac{\pi}{3}\right)\left(\cos\frac{\pi}{3}\right) + \sin\frac{\pi}{3}\right] \\ &= (0 + 1) - \left(-\frac{\pi}{6} + \frac{\sqrt{3}}{2}\right) \\ &= 1 + \frac{\pi}{6} - \frac{\sqrt{3}}{2} \\ &= \frac{6 - 3\sqrt{3} + \pi}{6}\end{aligned}$$

Example 5-38: Evaluate $\int_1^e x^2 \ln x\, dx.$

Using integration by parts with

$$u = \ln x \text{ and } dv = x^2\, dx$$
$$du = \frac{1}{x}\, dx \quad v = \frac{1}{3}x^3$$

you find that

$$\begin{aligned}\int x^2 \ln x\, dx &= \frac{1}{3}x^3 \ln x - \int \left(\frac{1}{3}x^3\right)\left(\frac{1}{x}\right) dx\\ &= \frac{1}{3}x^3 \ln x - \frac{1}{3}\int x^2\, dx\\ &= \frac{1}{3}x^3 \ln x - \frac{1}{3}\cdot\frac{1}{3}x^3 + C\\ &= \frac{1}{3}x^3 \ln x - \frac{1}{9}x^3 + C\end{aligned}$$

hence,

$$\begin{aligned}\int_1^e x^2 \ln x\, dx &= \left[\frac{1}{3}x^3 \ln x - \frac{1}{9}x^3\right]_1^e\\ &= \left[\frac{1}{3}(e)^3 \ln e - \frac{1}{9}(e)^3\right] - \left[\frac{1}{3}(1)^3 \ln 1 - \frac{1}{9}(1)^3\right]\\ &= \left[\frac{1}{3}(e)^3 - \frac{1}{9}(e)^3\right] - \left[0 - \frac{1}{9}\right]\\ &= \frac{2}{9}e^3 + \frac{1}{9}\\ &= \frac{1}{9}(2e^3 + 1)\end{aligned}$$

Example 5-39: Evaluate $\int_{\pi/4}^{\pi/2} \cot^4 x\, dx$

$$\begin{aligned}\int \cot^4 x\, dx &= \int \cot^2 x \cot^2 x\, dx\\ &= \int \cot^2 x(\csc^2 x - 1)\, dx\\ &= \int (\cot^2 x \csc^2 x - \cot^2 x)\, dx\\ &= \int \cot^2 x \csc^2 x\, dx - \int \cot^2 x\, dx\\ &= \int \cot^2 x \csc^2 x\, dx - \int (\csc^2 x - 1)\, dx\\ &= -\frac{1}{3}\cot^3 x + \cot x + x + C\end{aligned}$$

hence, $$\int_{\pi/4}^{\pi/2} \cot^4 x\,dx = \left[-\frac{1}{3}\cot^3 x + \cot x + x\right]_{\pi/4}^{\pi/2}$$
$$= \left[0 + 0 + \frac{\pi}{2}\right] - \left[-\frac{1}{3} + 1 + \frac{\pi}{4}\right]$$
$$= \frac{\pi}{4} - \frac{2}{3} \text{ or } \frac{3\pi - 8}{12}$$

Example 5-40: Evaluate $\int_{-\pi/2}^{3\pi/2} \cos^2 4x\,dx.$

$$\int \cos^2 4x\,dx = \int \left(\frac{1+\cos 8x}{2}\right)dx$$
$$= \frac{1}{2}x + \frac{1}{16}\sin 8x + C$$

hence, $$\int_{-\pi/2}^{3\pi/2} \cos^2 4x\,dx = \left[\frac{1}{2}x + \frac{1}{16}\sin 8x\right]_{-\pi/2}^{3\pi/2}$$
$$= \left[\frac{1}{2}\left(\frac{3\pi}{2}\right) + \frac{1}{16}\sin 12\pi\right] - \left[\frac{1}{2}\left(-\frac{\pi}{2}\right) + \frac{1}{16}\sin(-4\pi)\right]$$
$$= \left[\frac{3\pi}{4} + 0\right] - \left[-\frac{\pi}{4} + 0\right]$$
$$= \frac{4\pi}{4}$$
$$= \pi$$

Example 5-41: Evaluate $\int_{-3}^{3} \frac{dx}{x^2+9}$.

Because the integrand contains the form $a^2 + x^2$,

let $x = a \tan\theta = 3\tan\theta$

$dx = 3\sec^2\theta\,d\theta$

and $x^2 + 9 = 9\sec^2\theta$ (Figure 5-4)

Figure 5-4 Diagram for Example 5-41.

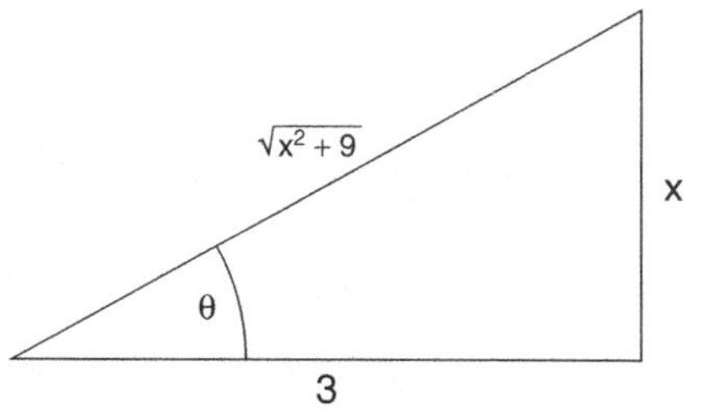

Hence, $\int \frac{dx}{x^2+9} = \int \frac{3\sec^2\theta d\theta}{9\sec^2\theta}$

$$= \frac{1}{3}\int d\theta$$

$$= \frac{1}{3}\theta + C$$

$$= \frac{1}{3}\tan^{-1}\frac{1}{3}x + C$$

and $\int_{-3}^{3} \frac{dx}{x^2+9} = \left[\frac{1}{3}\tan^{-1}\frac{1}{3}x\right]_{-3}^{3}$

$$= \frac{1}{3}\left[\tan^{-1}1 - \tan^{-1}(-1)\right]$$

$$= \frac{1}{3}\left[\frac{\pi}{4} - \left(-\frac{\pi}{4}\right)\right]$$

$$= \frac{1}{3}\left(\frac{\pi}{2}\right)$$

$$= \frac{\pi}{6}$$

Example 5-42: Evaluate $\int_{3}^{4} \frac{\sqrt{25-x^2}}{x}\,dx$.

Because the radical has the form $\sqrt{a^2-x^2}$,

let $x = a\sin\theta = 5\sin\theta$

$dx = 5\cos\theta\, d\theta$

and $\sqrt{25-x^2} = 5\cos\theta$ (Figure 5-5).

Figure 5-5 Diagram for Example 5-42.

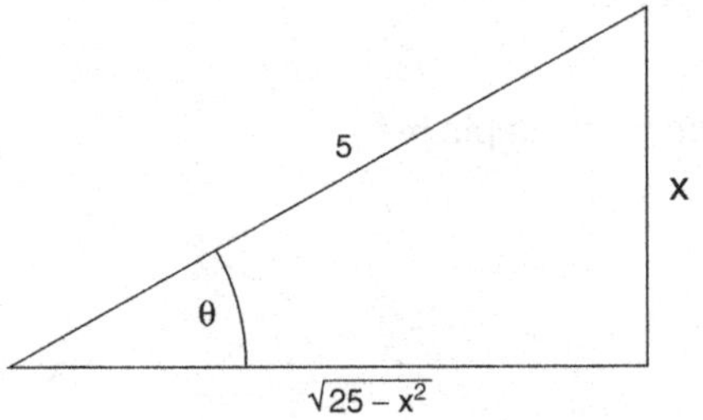

$$\text{Hence, } \int \frac{\sqrt{25-x^2}}{x}\,dx = \int \frac{5\cos\theta}{5\sin\theta}(5\cos\theta\, d\theta)$$

$$= 5\int \frac{\cos^2\theta}{\sin\theta}\,d\theta$$

$$= 5\int \frac{1-\sin^2\theta}{\sin\theta}\,d\theta$$

$$= 5\int (\csc\theta - \sin\theta)\,d\theta$$

$$= -5\ln|\csc\theta + \cot\theta| + 5\cos\theta + C$$

$$= -5\ln\left|\frac{5}{x} + \frac{\sqrt{25-x^2}}{x}\right| + 5\cdot\frac{\sqrt{25-x^2}}{5} + C$$

$$= -5\ln\left|\frac{5+\sqrt{25-x^2}}{x}\right| + \sqrt{25-x^2} + C$$

$$\text{and } \int_3^4 \frac{\sqrt{25-x^2}}{x}\,dx = \left[-5\ln\left|\frac{5+\sqrt{25-x^2}}{x}\right| + \sqrt{25-x^2}\right]_3^4$$

$$= [-5\ln 2 + 3] - [-5\ln 3 + 4]$$

$$= 5(\ln 3 - \ln 2) - 1$$

$$= 5\ln\frac{3}{2} - 1$$

Chapter Checkout

Q&A

1. Evaluate $\int \left(\frac{1}{x} + \sqrt{x} - e^x + \sin x\right) dx$.

2. Evaluate $\int \frac{x^2}{x^3+1}\,dx$.

3. Evaluate $\int xe^x\,dx$.

4. Evaluate $\int_0^{\pi/2} \sin x\cos^3 x\,dx$.

5. Evaluate $\int_1^2 \frac{\sqrt{4-x^2}}{x^2}\,dx$.

Answers: 1. $\ln|x| + 2/3\,x^{3/2} - e^x - \cos x + C$ **2.** $\ln|x^3+1|/3 + C$ **3.** $x e^x - e^x + C$ **4.** $1/4$ **5.** $\sqrt{3} - \pi/3$

Chapter 6

APPLICATIONS OF THE DEFINITE INTEGRAL

Chapter Check-In

- ❑ Calculating areas with definite integrals
- ❑ Finding volumes with definite integrals
- ❑ Computing arc lengths with definite integrals

The definite integral of a function has applications to many problems in calculus. Those considered in this chapter are areas bounded by curves, volumes by slicing, volumes of solids of revolution, and the lengths of arcs of a curve.

Area

The area of a region bounded by a graph of a function, the x-axis, and two vertical boundaries can be determined directly by evaluating a definite integral. If $f(x) \geq 0$ on $[a,b]$, then the area (A) of the region lying below the graph of $f(x)$, above the x-axis, and between the lines $x = a$ and $x = b$ is

$$A = \int_a^b f(x)\,dx$$

Figure 6-1 Finding the area under a non-negative function.

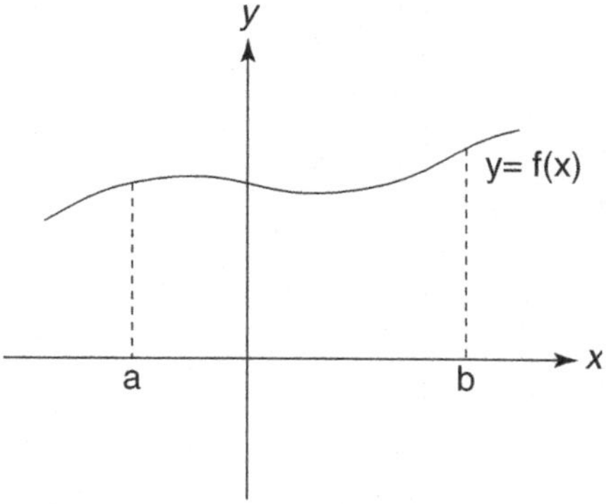

If $f(x) \leq 0$ on $[a,b]$, then the area (A) of the region lying above the graph of $f(x)$, below the x-axis, and between the lines $x = a$ and $x = b$ is

$$A = \left| \int_a^b f(x)\,dx \right|$$

$$\text{or} \quad A = -\int_a^b f(x)\,dx$$

Figure 6-2 Finding the area above a negative function.

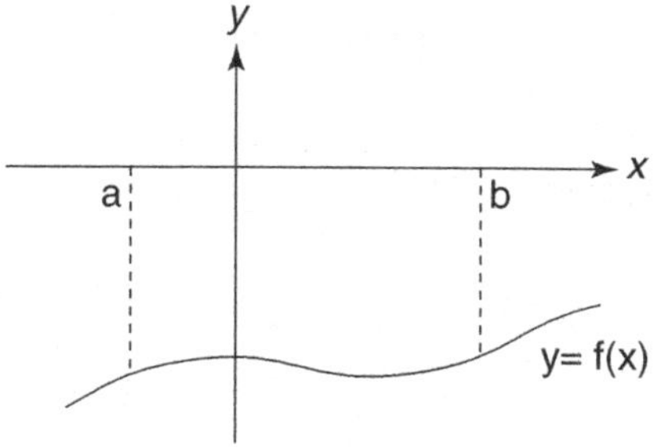

If $f(x) \geq 0$ on $[a,c]$ and $f(x) \leq 0$ on $[c,b]$, then the area (A) of the region bounded by the graph of $f(x)$, the x-axis, and the lines $x = a$ and $x = b$ would be determined by the following definite integrals:

$$A = \int_a^b |f(x)|\,dx$$

$$A = \int_a^c f(x)\,dx - \int_c^b f(x)\,dx$$

Figure 6-3 The area bounded by a function whose sign changes.

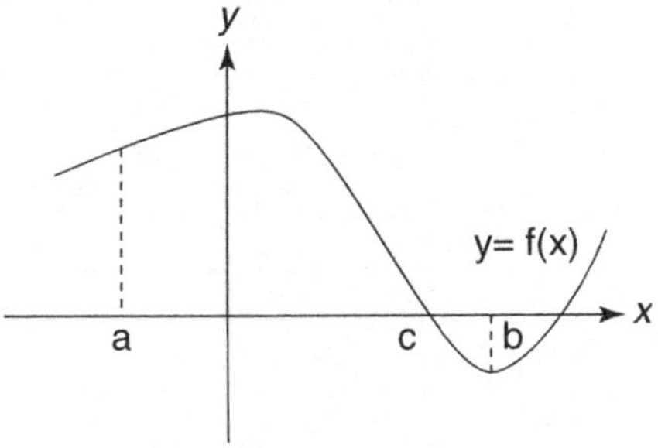

Note that in this situation it would be necessary to determine all points where the graph $f(x)$ crosses the x-axis and the sign of $f(x)$ on each corresponding interval.

For some problems that ask for the area of regions bounded by the graphs of two or more functions, it is necessary to determined the position of each graph relative to the graphs of the other functions of the region. The points of intersection of the graphs might need to be found in order to identify the limits of integration. As an example, if $f(x) \geq g(x)$ on $[a,b]$, then the area (A) of the region between the graphs of $f(x)$ and $g(x)$ and the lines $x = a$ and $x = b$ is

$$A = \int_a^b \left[f(x) - g(x)\right] dx$$

Figure 6-4 The area between two functions.

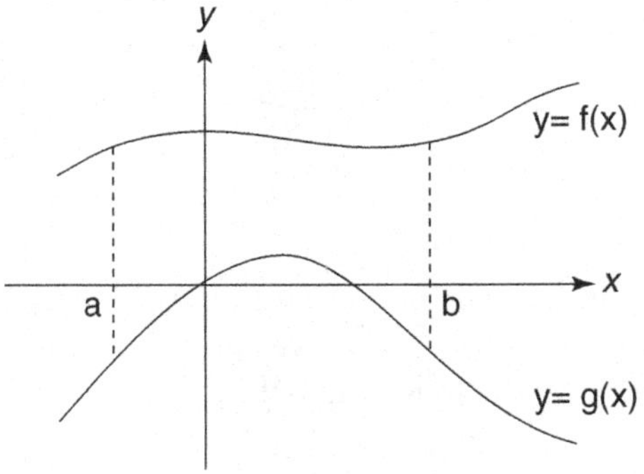

Note that an analogous discussion could be given for areas determined by graphs of functions of y, the y-axis, and the lines $y = a$ and $y = b$.

Example 6-1: Find the area of the region bounded by $y = x^2$, the x-axis, $x = -2$, and $x = 3$.

Because $f(x) \geq 0$ on $[-2,3]$, the area (A) is

$$\begin{aligned} A &= \int_{-2}^{3} x^2\, dx \\ &= \frac{1}{3} x^3 \Big]_{-2}^{3} \\ &= \frac{1}{3}(3)^3 - \frac{1}{3}(-2)^3 \\ A &= \frac{35}{3} \text{ or } 11\frac{2}{3} \end{aligned}$$

Example 6-2: Find the area of the region bounded by $y = x^3 + x^2 - 6x$ and the x-axis.

Setting $y = 0$ to determine where the graph intersects the x-axis, you find that

$$\begin{aligned} x^3 + x^2 - 6x &= 0 \\ x(x^2 + x - 6) &= 0 \\ x(x+3)(x-2) &= 0 \\ x = 0, x = -3&, x = 2 \end{aligned}$$

Because $f(x) \geq 0$ on $[-3,0]$ and $f(x) \leq 0$ on $[0,2]$ (see Figure 6-5), the area (A) of the region is

$$\begin{aligned} A &= \int_{-3}^{2} \left|(x^3 + x^2 - 6x)\right| dx \\ &= \int_{-3}^{0} (x^3 + x^2 - 6x)\, dx - \int_{0}^{2} (x^3 + x^2 - 6x)\, dx \\ &= \left[\frac{1}{4}x^4 + \frac{1}{3}x^3 - 3x^2\right]_{-3}^{0} - \left[\frac{1}{4}x^4 + \frac{1}{3}x^3 - 3x^2\right]_{0}^{2} \\ &= \frac{63}{4} - \left(-\frac{16}{3}\right) \\ A &= \frac{253}{12} \text{ or } 21\frac{1}{12} \end{aligned}$$

Figure 6-5 Diagram for Example 6-2.

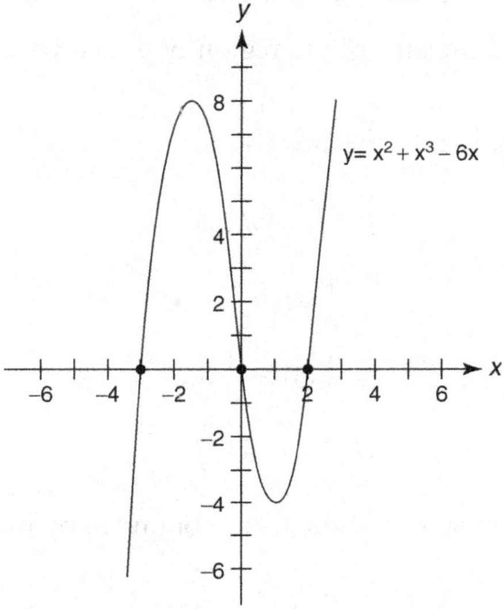

Example 6-3: Find the area bounded by $y = x^2$ and $y = 8 - x^2$.

Because $y = x^2$ and $y = 8 - x^2$, you find that

$$x^2 = 8 - x^2$$
$$2x^2 - 8 = 0$$
$$2(x^2 - 4) = 0$$
$$2(x + 2)(x - 2) = 0$$
$$x = -2, x = 2$$

hence, the curves intersect at (–2,4) and (2,4). Because $8 - x^2 \geq x^2$ on [–2,2] (see Figure 6-6), the area (A) of the region is

$$A = \int_{-2}^{2} \left[(8 - x^2) - (x^2)\right] dx$$
$$= \int_{-2}^{2} (8 - 2x^2)\, dx$$
$$= \left[8x - \frac{2}{3}x^3\right]_{-2}^{2}$$
$$= \frac{32}{3} - \left(-\frac{32}{3}\right)$$
$$A = \frac{64}{3} \text{ or } 21\frac{1}{3}$$

Figure 6-6 Diagram for Example 6-3.

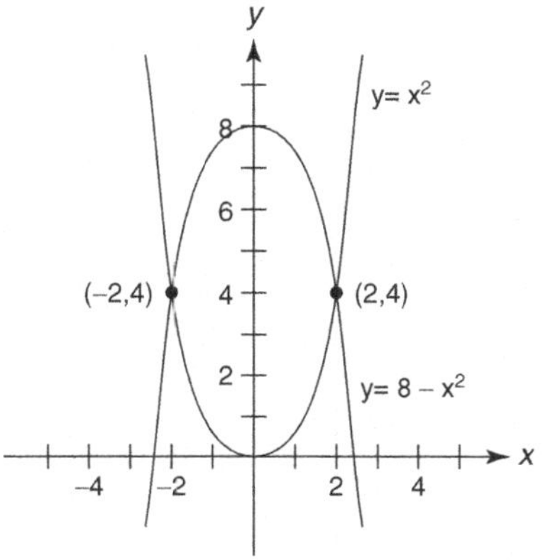

Volumes of Solids with Known Cross Sections 6.2/ET7.2

You can use the definite integral to find the volume of a solid with specific cross sections on an interval, provided you know a formula for the region determined by each cross section. If the cross sections generated are perpendicular to the x-axis, then their areas will be functions of x, denoted by $A(x)$. The volume (V) of the solid on the interval $[a,b]$ is

$$V = \int_a^b A(x)\,dx$$

If the cross sections are perpendicular to the y-axis, then their areas will be functions of y, denoted by $A(y)$. In this case, the volume (V) of the solid on $[a,b]$ is

$$V = \int_a^b A(y)\,dy$$

Example 6-4: Find the volume of the solid whose base is the region inside the circle $x^2 + y^2 = 9$ if cross sections taken perpendicular to the y-axis are squares.

Because the cross sections are squares perpendicular to the y-axis, the area of each cross section should be expressed as a function of y. The length of the side of the square is determined by two points on the circle $x^2 + y^2 = 9$ (Figure 6-7).

Figure 6-7 Diagram for Example 6-4.

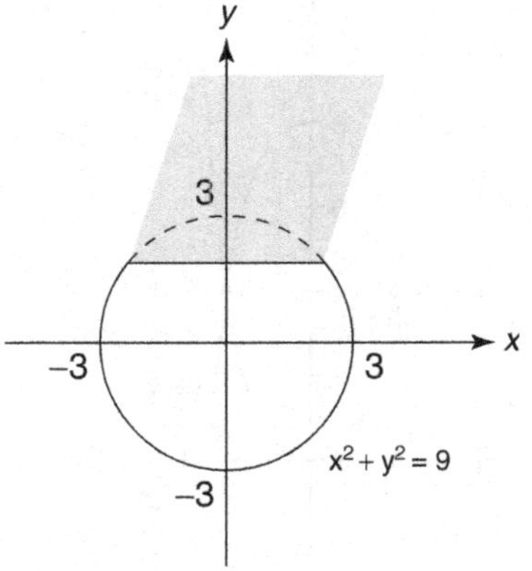

The area (A) of an arbitrary square cross section is $A = s^2$, where

$$s = 2\sqrt{9 - y^2}; \text{ hence,}$$

$$A(y) = \left[2\sqrt{9 - y^2}\right]^2$$
$$A(y) = 4(9 - y^2)$$

The volume (V) of the solid is

$$V = \int_{-3}^{3} 4(9 - y^2)\,dy$$
$$= 4\left[9y - \frac{1}{3}y^3\right]_{-3}^{3}$$
$$= 4[18 - (-18)]$$
$$V = 144$$

Example 6-5: Find the volume of the solid whose base is the region bounded by the lines $x + 4y = 4$, $x = 0$, and $y = 0$, if the cross sections taken perpendicular to the x-axis are semicircles.

Because the cross sections are semicircles perpendicular to the x-axis, the area of each cross section should be expressed as a function of x. The diameter of the semicircle is determined by a point on the line $x + 4y = 4$ and a point on the x-axis (Figure 6-8).

Figure 6-8 Diagram for Example 6-5.

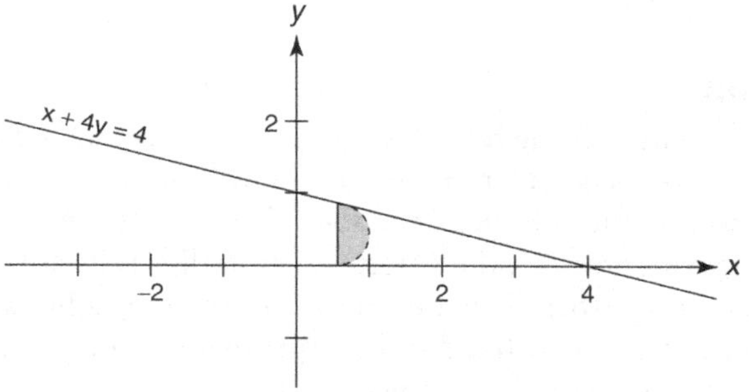

The area (A) of an arbitrary semicircle cross section is

$$A = \frac{1}{2}\pi r^2 = \frac{1}{2}\pi\left(\frac{1}{2}d\right)^2$$

where $$d = \frac{4-x}{4} \text{ and } r = \frac{4-x}{8}$$

hence, $$A(x) = \frac{1}{2}\pi\left(\frac{4-x}{8}\right)^2$$

$$A(x) = \frac{1}{128}\pi(4-x)^2$$

The volume (V) of the solid is

$$\begin{aligned}
V &= \int_0^4 \frac{1}{128}\pi(4-x)^2\,dx \\
&= \frac{1}{128}\pi\int_0^4 (16 - 8x + x^2)\,dx \\
&= \frac{1}{128}\pi\left[16x - 4x^2 + \frac{1}{3}x^3\right]_0^4 \\
&= \frac{1}{128}\pi\left[\frac{64}{3}\right] \\
V &= \frac{\pi}{6}
\end{aligned}$$

Volumes of Solids of Revolution

You can also use the definite integral to find the volume of a solid that is obtained by revolving a plane region about a horizontal or vertical line that does not pass through the plane. This type of solid will be made up of one of three types of elements—disks, washers, or cylindrical shells—each of which requires a different approach in setting up the definite integral to determine its volume.

Disk method

If the axis of revolution is the boundary of the plane region and the cross sections are taken perpendicular to the axis of revolution, then you use the **disk method** to find the volume of the solid. Because the cross section of a disk is a circle with area πr^2, the volume of each disk is its area times its thickness. If a disk is perpendicular to the x-axis, then its radius should be expressed as a function of x. If a disk is perpendicular to the y-axis, then its radius should be expressed as a function of y.

The volume (V) of a solid generated by revolving the region bounded by $y = f(x)$ and the x-axis on the interval $[a,b]$ about the x-axis is

$$V = \int_a^b \pi \left[f(x) \right]^2 dx$$

If the region bounded by $x = f(y)$ and the y-axis on $[a,b]$ is revolved about the y-axis, then its volume (V) is

$$V = \int_a^b \pi \left[f(y) \right]^2 dy$$

Note that $f(x)$ and $f(y)$ represent the radii of the disks or the distance between a point on the curve to the axis of revolution.

Example 6-6: Find the volume of the solid generated by revolving the region bounded by $y = x^2$ and the x-axis on $[2,3]$ about the x-axis.

Because the x-axis is a boundary of the region, you can use the disk method (see Figure 6-9).

Figure 6-9 Diagram for Example 6-6.

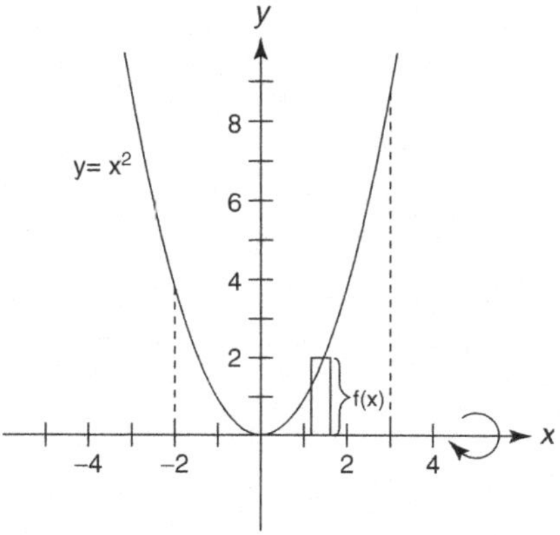

The volume (V) of the solid is

$$V = \int_{-2}^{3} \pi (x^2)^2 \, dx$$
$$= \pi \int_{-2}^{3} x^4 \, dx$$
$$= \pi \left[\frac{1}{5} x^5 \right]_{-2}^{3}$$
$$= \pi \left[\frac{243}{5} - \left(\frac{-32}{5} \right) \right]$$
$$V = 55\pi$$

Washer method

If the axis of revolution is not a boundary of the plane region and the cross sections are taken perpendicular to the axis of revolution, you use the **washer method** to find the volume of the solid. Think of the washer as a "disk with a hole in it" or as a "disk with a disk removed from its center." If R is the radius of the outer disk and r is the radius of the inner disk, then the area of the washer is $\pi R^2 - \pi r^2$, and its volume would be its area times its thickness. As noted in the discussion of the disk method, if a washer is perpendicular to the x-axis, then the inner and outer radii should be expressed as functions of x. If a washer is perpendicular to the y-axis, then the radii should be expressed as functions of y.

The volume (V) of a solid generated by revolving the region bounded by $y = f(x)$ and $y = g(x)$ on the interval $[a,b]$ where $f(x) \geq g(x)$, about the x-axis is

$$V = \int_a^b \pi \left\{ [f(x)]^2 - [g(x)]^2 \right\} dx$$

If the region bounded by $x = f(y)$ and $x = g(y)$ on $[a,b]$, where $f(y) \geq g(y)$ is revolved about the y-axis, then its volume (V) is

$$V = \int_a^b \pi \left\{ [f(y)]^2 - [g(y)]^2 \right\} dy$$

Note again that $f(x)$ and $g(x)$ and $f(y)$ and $g(y)$ represent the outer and inner radii of the washers or the distance between a point on each curve to the axis of revolution.

Example 6-7: Find the volume of the solid generated by revolving the region bounded by $y = x^2 + 2$ and $y = x + 4$ about the x-axis.

Because $y = x^2 + 2$ and $y = x + 4$, you find that

$$x^2 + 2 = x + 4$$
$$x^2 - x - 2 = 0$$
$$(x + 1)(x - 2) = 0$$
$$x = -1, x = 2$$

The graphs will intersect at $(-1,3)$ and $(2,6)$ with $x + 4 \geq x^2 + 2$ on $[-1,2]$ (Figure 6-10).

Figure 6-10 Diagram for Example 6-7.

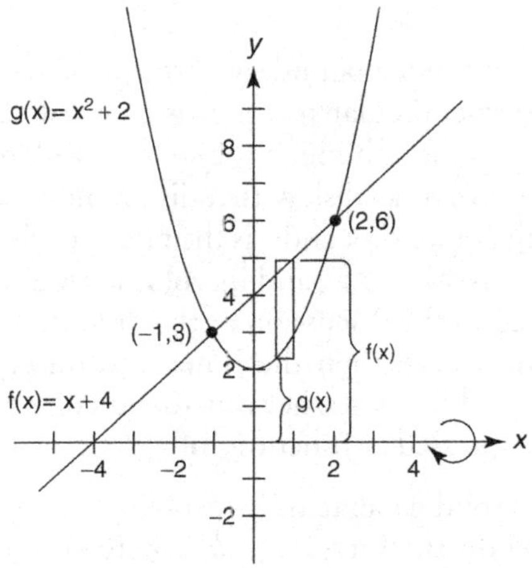

Because the x-axis is not a boundary of the region, you can use the washer method, and the volume (V) of the solid is

$$V = \int_{-1}^{2} \pi\left[(x+4)^2 - \left(x^2+2\right)^2\right]dx$$
$$= \int_{-1}^{2} \pi\left[\left(x^2+8x+16\right) - \left(x^4+4x^2+4\right)\right]dx$$
$$= \pi \int_{-1}^{2} \left(-x^4 - 3x^2 + 8x + 12\right)dx$$
$$= \pi\left[-\tfrac{1}{5}x^5 - x^3 + 4x^2 + 12x\right]_{-1}^{2}$$
$$= \pi\left[\frac{128}{5} - \left(-\frac{34}{5}\right)\right]$$
$$V = \frac{162\pi}{5}$$

Cylindrical shell method

If the cross sections of the solid are taken parallel to the axis of revolution, then the **cylindrical shell method** will be used to find the volume of the solid. If the cylindrical shell has radius r and height h, then its volume would be $2\pi rh$ times its thickness. Think of the first part of this product, ($2\pi rh$), as the area of the rectangle formed by cutting the shell perpendicular to its radius and laying it out flat. If the axis of revolution is vertical, then the radius and height should be expressed in terms of x. If, however, the axis of revolution is horizontal, then the radius and height should be expressed in terms of y.

The volume (V) of a solid generated by revolving the region bounded by $y = f(x)$ and the x-axis on the interval $[a, b]$, where $f(x) \geq 0$, about the y-axis is

$$V = \int_a^b 2\pi x\, f(x)\, dx$$

If the region bounded by $x = f(y)$ and the y-axis on the interval $[a, b]$, where $f(y) \geq 0$, is revolved about the x-axis, then its volume (V) is

$$V = \int_a^b 2\pi y\, f(y)\, dy$$

Note that the x and y in the integrands represent the radii of the cylindrical shells or the distance between the cylindrical shell and the axis of revolution. The $f(x)$ and $f(y)$ factors represent the heights of the cylindrical shells.

Example 6-8: Find the volume of the solid generated by revolving the region bounded by $y = x^2$ and the x-axis [1,3] about the y-axis.

In using the cylindrical shell method, the integral should be expressed in terms of x because the axis of revolution is vertical. The radius of the shell is x, and the height of the shell is $f(x) = x^2$ (Figure 6-11).

Figure 6-11 Diagram for Example 6-8.

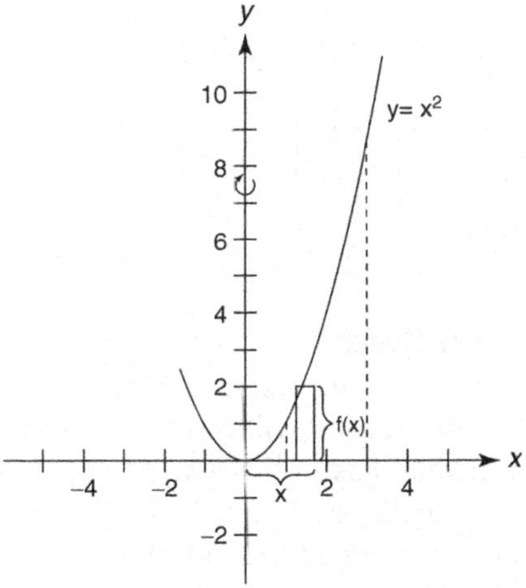

The volume (V) of the solid is

$$V = \int_1^3 2\pi x \cdot x^2\,dx$$

$$= 2\pi \int_1^3 x^3\,dx$$

$$= 2\pi \left[\frac{1}{4}x^4\right]_1^3$$

$$= \frac{1}{2}\pi\,(81 - 1)$$

$$V = 40\pi$$

Arc Length

The length of an arc along a portion of a curve is another application of the definite integral. The function and its derivative must both be continuous on the closed interval being considered for such an arc length to be guaranteed. If $y = f(x)$ and $y' = f'(x)$ are continuous on the closed interval $[a,b]$, then the arc length (L) of $f(x)$ on $[a,b]$ is

$$L = \int_a^b \sqrt{1 + [f'(x)]^2}\, dx$$

Similarly, if $x = f(y)$ and $x' = f'(y)$ are continuous on the closed interval $[a,b]$, then the arc length (L) of $f(y)$ on $[a,b]$ is

$$L = \int_a^b \sqrt{1 + [f'(y)]^2}\, dy$$

Example 6-9: Find the arc length of the graph of $f(x) = \frac{1}{3}x^{3/2}$ on the interval $[0,5]$.

Because

$$f(x) = \frac{1}{3}x^{3/2}$$

$$f'(x) = \frac{1}{2}x^{1/2}$$

and

$$\begin{aligned} L &= \int_0^5 \sqrt{1 + \left(\frac{1}{2}x^{1/2}\right)^2}\, dx \\ &= \int_0^5 \sqrt{1 + \frac{1}{4}x}\, dx \\ &= \int_0^5 \left(1 + \frac{1}{4}x\right)^{1/2} dx \\ &= \left[4 \cdot \frac{2}{3}\left(1 + \frac{1}{4}x\right)^{3/2}\right]_0^5 \\ &= \frac{8}{3}\left(\frac{27}{8} - 1\right) \\ L &= \frac{19}{3} \end{aligned}$$

Example 6-10: Find the arc length of the graph of $f(x) = \ln(\sin x)$ on the interval $[\pi/4, \pi/2]$.

Because

$$f(x) = \ln(\sin x)$$

$$f'(x) = \frac{\cos x}{\sin x} = \cot x$$

and

$$\begin{aligned} L &= \int_{\pi/4}^{\pi/2} \sqrt{1 + \cot^2 x}\, dx \\ &= \int_{\pi/4}^{\pi/2} \sqrt{\csc^2 x}\, dx \\ &= \int_{\pi/4}^{\pi/2} \csc x\, dx \\ &= \left[-\ln\left|\csc x + \cot x\right|\right]_{\pi/4}^{\pi/2} \\ &= (-\ln 1) - \left(-\ln\left|\sqrt{2} + 1\right|\right) \\ &= 0 + \ln\left|\sqrt{2} + 1\right| \\ L &= \ln\left|\sqrt{2} + 1\right| \\ L &\approx 0.8813736 \end{aligned}$$

Chapter Checkout

Q&A

1. Find the area of the region bounded by $y = x^3 - 25x$ and the x-axis.
2. Find the area of the region bounded by $y = x^2$ and $y = x$.
3. Find the volume of the solid whose base is the region bounded by the lines $x + 5y = 5$, $x = 0$, and $y = 0$, if cross sections taken perpendicular to the x-axis are squares.

4. Find the volume of the solid generated by revolving the region bounded by $y = x^2$ and $y = x$ about the x-axis.
5. Find the arc length of the graph of $f(x) = \ln(\sec x)$ on the interval $[0, \pi/6]$.

Answers: 1. 625/2 **2.** 1/6 **3.** 5/3 **4.** $2\pi/15$ **5.** $(\ln 3)/2$

CQR REVIEW

Use this CQR Review to practice what you've learned in this book. After you work through the review questions, you're well on your way to achieving your goal of understanding calculus.

Chapter 1

1. Which of the following can **not** be used to find the slope of a line?

   a. m = rise / run
   b. $m = (y_1 - x_1) / (y_2 - x_2)$
   c. $m = (y_1 - y_2) / (x_1 - x_2)$
   d. $m = (y_2 - y_1) / (x_2 - x_1)$

2. Which of the following are functions?

   a. $f(x) = 5x + 3$
   b. $y = \cos 5x$
   c. $x = 2$
   d. $f(x) = 6/(x^2 + 4)$
   e. $x^2 + y^2 = 144$

3. Any nonvertical lines are parallel if they have ______.

4. Two nonvertical, nonhorizontal lines are perpendicular if the product of their slopes is ______.

5. Find an equation of the line that has slope 6/5 and crosses the y-axis at 3.

6. Complete the trigonometric identity for the following:

   a. $\cos(x - y)$
   b. $1/\cot x$
   c. $\cos(-x)$
   d. $\cos x/\sin x$
   e. $\csc^2 x - 1$
   f. $\sin^2 x + \cos^2 x$

Chapter 2

7. Evaluate $\lim_{x \to 1} \frac{x^3 - 1}{x - 1}$.

8. Evaluate $\lim_{x \to +\infty} \frac{1 - x^3}{3x^3 + x^2 - 1}$.

Chapter 3

9. In addition to the notation $f'(x)$, which the following can be used to represent the derivative of $y = f(x)$?

a. Df_x
b. $df(x)/dx$
c. y'
d. dx/dy

10. Complete the following statements about trigonometric function differentiation.

a. If $f(x) = \sin x$, then $f'(x) =$
b. If $f(x) = \cos x$, then $f'(x) =$
c. If $f(x) = \tan x$, then $f'(x) =$
d. If $f(x) = \cot x$, then $f'(x) =$
e. If $f(x) = \sec x$, then $f'(x) =$
f. If $f(x) = \csc x$, then $f'(x) =$

11. Find y' if $y = \sqrt[3]{\sin x + \pi}$.

12. Find $f'(x)$ if $f(x) = \sqrt{x^2 + 1}$.

Chapter 4

13. The ______ is the line that is perpendicular to the tangent line at the point of tangency.

14. The point $(x, f(x))$ is called a critical point of $f(x)$ if x is in the ______ of the function, and either $f'(x) =$ ______ or ______.

15. If the derivative of a function is greater than zero at each point on an interval I, then the function is said to be ______ on I. If the derivative of a function is less than zero at each point on an interval I, then the function is said to be ______ on I.

16. You can **not** use the Second Derivative Test for Relative Extrema in which of the following situations?

a. $f'(x) = 0$ and $f''(x) = 0$
b. $f'(x) = 0$ and $f''(x)$ does not exist
c. $f'(x) = -(f''(x))$
d. $f'(x)$ does not exist

17. Find the equation of the tangent line to the graph of $f(x) = x^2e^{-x}$ at the point $(1,1/e)$.

18. Find the maximum and minimum values of $f(x) = x^3 - 3x^2 - 9x + 4$ on the interval $[-2,6]$.

19. Find the equation of the tangent line to the graph of $f(x) = \sin(x^2)$ at the point $(0,0)$.

20. Find the maximum value of $y = \frac{x}{x^2 + 1}$ on the interval $[0,5]$.

21. Water is dripping into a cylindrical can with a radius of 3 inches. If the volume is increasing at a rate of 2 cubic inches per minute, how fast is the depth changing?

22. Two cars are traveling toward an intersection, one heading north at a rate of 65 mph and the second heading west at a rate of 45 mph. Find the rate of change of the distance between the two cars when both are 1 mile from the intersection.

23. If you know that a function is increasing on the interval $(0,3)$ and decreasing on the interval $(3,6)$, does this imply that the function has a local maximum when $x = 3$? What sorts of situations are possible?

24. What's wrong with the problem "A rectangular box is to have a volume of 8 square units. Find the maximum surface area such a box could have."

Chapter 5

25. What is the acceleration of a free-falling object due to gravity?

26. According to the Fundamental Theorem of Calculus: The value of the definite integral of a function on $[a,b]$ is the ______ of any ______ evaluated at the upper limit of integration and the same antiderivative evaluated at the lower limit of integration.

27. Evaluate $\int \cos^6 x \sin^3 x\, dx$.

28. Evaluate $\int_0^1 xe^x\, dx$.

29. Evaluate $\int \frac{x}{\sqrt{4-x^2}}\, dx$.

30. Evaluate $\int_1^2 x \ln x\, dx$.

31. A rock is thrown upward from a 200 foot cliff with an initial velocity of 30 feet per second. How long will it take for the ball to hit the ground?

32. What's wrong with the computation $\int_{-1}^{1} \frac{1}{x^2}\, dx = -1x^{-1}\Big|_{-1}^{1} = \left(-\frac{1}{1}\right) - \left(-\frac{1}{-1}\right) = -2$?

33. If a highway patrol officer is sitting off to the side of a road monitoring the speed of approaching traffic, does it matter how far off the roadway the officer sits? (Hint: Think about it as a related rates problem with one of the rates of change being zero.)

34. Fast cars are often rated for how quickly they can accelerate from 0 to 60 miles per hour (which is equivalent to 88 feet per second). If a car takes s seconds to accelerate from 0 to 60 (and supposing constant acceleration), over what distance will it travel in the process?

Chapter 6

35. Find the area of the region bounded by $y = x$ and $y = x^3$.

36. Find the area of the region bounded between $y = x^2$ and $y = \sqrt{x}$.

37. Find the volume of the solid generated by revolving the region bounded by $y = \sin x$ and the x-axis on $[0, \pi]$ about the y-axis.

38. Find the arc length of the line $y = 2x$ on the interval $[0,3]$.

39. Find the volume of the solid generated by revolving the region bounded by $y = 1/x$ and the x-axis on $[1, 20]$ about the x-axis.

40. The area bounded by the function $f(x)$ and the x-axis between $x = a$ and $x = b$ is given by $\int_a^b |f(x)|\, dx$. Is this the same as $\left|\int_a^b f(x)\, dx\right|$?

41. One of the early practical problems to which calculus was applied was determining the volume of a barrel that was filled with liquid when it was impractical to just pour the liquid out for measuring. Select an object with round cross sections and see how accurately you can find its volume by treating it as the solid of revolution generated by some curve.

42. Find the dimensions of a cylinder with a volume of 100 cubic units which minimize surface area. If cost of materials were the only consideration, presumably this would be the shape all canned goods would be sold in, but of course a trip to the grocery store makes it clear that few cans are this shape. Why would this be so, and what other factors influence the shapes of different cylindrical containers? Compare your ideas to the dimensions of actual cans to see how well they agree.

Answers: 1. b **2.** a, b, and d **3.** the same slope **4.** –1 **5.** $6x - 5y = -15$ **6.** (a) $\cos x \cos y - \sin x \sin y$ (b) $\tan x$ (c) $\cos x$ (d) $\cot x$ (e) $\cot^2 x$ (f) 1 **7.** 3 **8.** –1/3 **9.** b and c **10.** (a) $\cos x$ (b) $-\sin x$ (c) $\sec^2 x$ (d) $-\csc^2 x$ (e) $\sec x \tan x$ (f) $-\csc x \cot x$ **11.** $\frac{1}{3}(\sin x + \pi)$ **12.** $\frac{x}{\sqrt{x^2+1}}$ **13.** normal line **14.** domain, 0, does not exist **15.** increasing, decreasing **16.** a, b, and d **17.** $\frac{1}{e}x - y = 0$ **18.** minimum –23, maximum 58 **19.** $y = 0$ **20.** 1/2 **21.** $\frac{2}{9\pi}$ inches per minute **22.** $-110/\sqrt{2}$ **23.** Provide your own answer **24.** Provide your own answer **25.** –32 ft/sec^2 **26.** difference, antiderivative of the function **27.** $\frac{1}{9}\cos^9 x - \frac{1}{7}\cos^7 x + C$ **28.** 1 **29.** $-\sqrt{4-x^2}$ **30.** 2 ln 2 – 3/4 **31.** approximately 4.6 seconds **32.** Provide your own answer **33.** Yes, the greater the distance off the road the lower an approaching car's speed is relative to the officer. **34.** 44s feet **35.** 1/2 **36.** 1/3 **37.** $2\pi^2$ **38.** $3\sqrt{5}$ **39.** $\frac{19\pi}{20}$ **40.** Provide your own answer **41.** Provide your own answer **42.** Provide your own answer

CQR RESOURCE CENTER

CQR Resource Center offers the best resources available in print and online to help you study and review the core concepts of calculus. You can find additional resources, plus study tips and tools to help test your knowledge, at `www.cliffsnotes.com`.

Books

This CliffsQuickReview book is just what it's called, a quick review of calculus. If you need to brush up more of the pre-requisites, or if you want a fuller discussion or other practical advice, check out these other publications:

CliffsQuickReview Basic Math and Pre-Algebra, by Jerry Bobrow, gives you a review of topics including the basics of working with fractions, decimals, powers, exponents, roots, and an introduction to algebraic expressions and solving equations. Wiley Publishing, Inc., 2001.

CliffsQuickReview Algebra I, by Jerry Bobrow, gives you a review of topics including sets, equations, polynomials, factoring, inequalities, graphing, and functions. Wiley Publishing, Inc., 2001.

CliffsQuickReview Geometry, by Edward Kohn and David Herzog, gives you a review of topics including perimeter, area, volume, the Pythagorean theorem, 30°-60°-90° and 45°-45°-90° triangles, and the basics of coordinate geometry including plotting points, distances, midpoints, slopes and equations of lines. Wiley Publishing, Inc., 2001.

CliffsQuickReview Algebra II, by Edward Kohn, gives you a review of topics including solving systems of equations, polynomials, factoring, complex numbers, conic sections, exponential and logarithmic functions, sequences and series, and other material pre-requisite for calculus. Wiley Publishing, Inc., 2001.

CliffsQuickReview Trigonometry, by David A. Kay, gives you a review of triangles, trigonometric functions and identities, vectors, polar coordinates, complex numbers, and inverse functions. Wiley Publishing, Inc., 2001.

CliffsQuickReview Linear Algebra, by Steven A. Leduc, is an in-depth look at algebraic equations and inequalities. Wiley Publishing, Inc., 1986.

CliffsAP Calculus AB and BC Preparation Guide, by Kerry King, gives you tips and suggestions for getting the most credit you can on the Advanced Placement Calculus AB and BC tests. The book reviews crucial calculus topics, introduces test-taking strategies, and includes sample questions and tests. Wiley Publishing, Inc., 2001.

Cliffs Math Review for Standardized Tests, by Jerry Bobrow, helps you to review, refresh, and prepare for standardized math tests. Each topic-specific review section includes a diagnostic test, rules and key concepts, practice problems, a review test, glossary, and a section devoted to key strategies, practice, and analysis for the most common types of standardized questions. Wiley Publishing, Inc., 1985.

How to Ace Calculus: The Streetwise Guide, by Joel Hass, Abigail Thompson, and Colin Conrad Adams, gives a lot of practical tips not just on the subject matter itself, but also on picking teachers and preparing for tests. W H Freeman & Co., 1998.

3000 Solved Problems in Calculus, by Elliot Mendelson, can give you all the extra practice problems you want. McGraw-Hill, 1992.

A Tour of the Calculus, by David Berlinski, gives a complete exploration of what many of the theorems of calculus really mean and a look at how the discipline of calculus is one of the human intellect's most impressive accomplishments. Vintage Books, 1997.

The Story of Mathematics, by Richard Mankiewicz and Ian Stewart, gives a very accessible account of the development of mathematics, including calculus, from the earliest archeological evidence on. Princeton University Press, 2001.

Wiley also has three Web sites that you can visit to read about all the books we publish:

- `www.cliffsnotes.com`
- `www.dummies.com`
- `www.wiley.com`

Internet

Visit the following Web sites for more information about calculus:

Ask Dr. Math—forum.swarthmore.edu/dr.math—is an award-winning site that offers a free question-and-answer service, as well as archives of past questions and answers.

Karl's Calculus Tutor—www.netsrq.com/~hahn/calc.html—is a complete calculus help site with entertaining and understandable explanations of most topics, free help with math problems, good links, and recommended books.

S.O.S. Mathematics—www.sosmath.com—is a nice site with a broad range of helpful pages covering algebra through calculus and beyond, including some animated graphics to demonstrate specific calculus ideas and some sample exams (with solutions).

calculus@internet—www.calculus.net—is an organized clearinghouse of links to a ton of other pages about math topics.

Visual Calculus—http://archives.math.utk.edu/visual.calculus/—is an award-winning Web site from the University of Tennessee that offers a wide variety of step-by-step illustrated tutorials on calculus topics including pre-calculus, limits, continuity, derivatives, integration, and sequences and series.

The MathServ Calculus Toolkit—http://mss.math.vanderbilt.edu/%7epscrooke/toolkit.shtml—is not the most graphically exciting Web site out there, but it does offer easy-to-use online programs that do the heavy lifting for you—everything from graphing functions and equations to computing limits.

The BHS Calculus Project—http://www.bhs-ms.org/calculus.htm—serves as an archive of student projects that show calculus's connection to the real world. Student research and reporting shows how calculus impacts everyday topics such as fractals, ice cones, bicycles, tape decks, and AIDS.

AP Calculus Problem of the Week—http://www.seresc.k12.nh.us/www/alvirne.html—offers a different calculus based problem every week. Visitors can also submit their own calculus challenges for future inclusion on the Web site.

Mathematica Animations—http://www.calculus.org/Contributions/animations.html—features short QuickTime movies that illustrate key calculus concepts such as the definition of a derivative, the second derivative function, the volume of cones, and Reimann sums.

Math for Morons Like Us: Pre-Calculus & Calculus—http://library.thinkquest.org/20991/calc/index.html—outlines major calculus topics as well as issues in other branches of mathematics. A fairly active pre-calculus and calculus message board enables visitors to ask and answer thought-provoking questions.

Help With Calculus For Idiots (Like Me)—ccwf.cc.utexas.edu/~egumtow/calculus—is another page with explanations of several calculus topics that gives practical advice about what you'll really need to know to get through a calculus class.

The Integrator—integrals.wolfram.com—actually computes integrals for you in the blink of an eye.

A History of the Calculus—www-history.mcs.st-and.ac.uk/history/HistTopics/The_rise_of_calculus.html—gives a good yet very brief survey of the origins of many of the major parts of modern calculus.

Next time you're on the Internet, don't forget to drop by www.cliffsnotes.com. We created an online Resource Center that you can use today, tomorrow, and beyond.

Glossary

antiderivative A function $F(x)$ is called an antiderivative of a function $f(x)$ if $F'(x) = f(x)$ for all x in the domain of f. In words, this means that an antiderivative of f is a function which has f for its derivative.

chain rule The chain rule tells how to find the derivative of composite functions. In symbols, the chain rule says $\frac{d}{dx}\left(f\left(g\left(x\right)\right)\right) = f'\left(g\left(x\right)\right) \cdot g'\left(x\right)$. In words, the chain rule says the derivative of a composite function is the derivative of the outside function, done to the inside function, times the derivative of the inside function.

change of variables A term sometimes used for the technique of integration by substitution.

concave down A function is concave down on an interval if $f'(x)$ is decreasing on that interval.

concave up A function is concave up on an interval if $f'(x)$ is increasing on that interval.

continuous A function $f(x)$ is continuous at a point $x = c$ when $f(c)$ is defined, $\lim_{x \to c} f(x)$ exists, and $\lim_{x \to c} f(x) = f(c)$. In words, this means the curve could be drawn without lifting the pencil. To say that a function is continuous on some interval means that it is continuous at each point in that interval.

critical point A critical point of a function is a point $(x, f(x))$ with x in the domain of the function and either $f'(x) = 0$ or $f'(x)$ undefined. Critical points are among the candidates to be maximum or minimum values of a function.

cylindrical shell method A procedure for finding the volume of a solid of revolution by treating it as a collection of nested thin rings.

definite integral The definite integral of $f(x)$ between $x = a$ and $x = b$, denoted $\int_a^b f(x)\,dx$, gives the signed area between $f(x)$ and the x-axis from $x = a$ to $x = b$, with area above the x-axis counting positive and area below the x-axis counting negative.

derivative The derivative of a function $f(x)$ is a function that gives the slope of $f(x)$ at each value of x. The derivative is most often denoted $f'(x)$ or $\frac{d}{dx}$. The mathematical definition of the derivative is $\lim_{w \to x} \frac{f(w) - f(x)}{w - x}$, or in words, the limit of the slopes of the secant lines through the point $(x, f(x))$ and a second point on the graph of $f(x)$ as that second point approaches the first. The derivative can be interpreted as the slope of a line tangent to the function, the instantaneous velocity of the function, or the instantaneous rate of change of the function.

differentiable A function is said to be differentiable at a point when the function's derivative exists at that point. A function will fail to be differentiable at places where the function is not continuous or where the function has corners.

disk method A procedure for finding the volume of a solid of revolution by treating it as a collection of thin slices with circular cross sections.

double intercept form The intercept form for the equation of a line is $x/a + y/b = 1$, where the line has its x-intercept (the place where the line crosses the x-axis) at the point $(a,0)$ and its y-intercept (the place where the line crosses the y-axis) at the point $(0,b)$.

Extreme Value Theorem A theorem stating that a function which is continuous on a closed interval $[a, b]$ must have a maximum and a minimum value on $[a, b]$.

First Derivative Test for Relative Extrema A method used to determine whether a critical point of a function is a relative maximum or relative minimum. If a continuous function changes from increasing (first derivative positive) to decreasing (first derivative negative) at a point, then that point is a relative maximum. If a function changes from decreasing (first derivative negative) to increasing (first derivative positive) at a point, then that point is a relative minimum.

general antiderivative If $F(x)$ is an antiderivative of a function $f(x)$, then $F(x) + C$ is called the general antiderivative of $f(x)$.

general form The general form (sometimes also called standard form) for the equation of a line is $ax + by = c$, where a and b are not both zero.

higher order derivatives The second derivative, third derivative, and so forth for some function.

implicit differentiation A procedure for finding the derivative of a function which has not been given explicitly in the form "$f(x) =$".

indefinite integral The indefinite integral of $f(x)$ is another term for the general antiderivative of $f(x)$. The indefinite integral of $f(x)$ is represented in symbols as $\int f(x)\,dx$.

instantaneous rate of change One way of interpreting the derivative of a function is to understand it as the instantaneous rate of change of that function, the limit of the average rates of change between a fixed point and other points on the curve that get closer and closer to the fixed point.

instantaneous velocity One way of interpreting the derivative of a function $s(t)$ is to understand it as the velocity at a given moment t of an object whose position is given by the function $s(t)$.

integration by parts One of the most common techniques of integration, used to reduce complicated integrals into one of the basic integration forms.

limit A function $f(x)$ has the value L for its limit as x approaches c if as the value of x gets closer and closer to c, the value of $f(x)$ gets closer and closer to L.

Mean Value Theorem If a function $f(x)$ is continuous on a closed interval $[a,b]$ and differentiable on the open interval (a,b), then there exists some c in the interval $[a,b]$ for which $f'(c)=\frac{f(b)-f(a)}{b-a}$.

normal line The normal line to a curve at a point is the line perpendicular to the tangent line at that point.

point of inflection A point is called a point of inflection of a function if the function changes from concave upward to concave downward, or vice versa, at that point.

point-slope form The point-slope form for the equation of a line is $y - y_1 = m(x - x_1)$, where m stands for the slope of the line and (x_1,y_1) is a point on the line.

Riemann sum A Riemann sum is a sum of several terms, each of the form $f(x_i)\Delta x$, each representing the area below a function $f(x)$ on some interval if $f(x)$ is positive or the negative of that area if $f(x)$ is negative. The definite integral is mathematically defined to be the limit of such a Riemann sum as the number of terms approaches infinity.

Second Derivative Test for Relative Extrema A method used to determine whether a critical point of a function is a relative maximum or relative minimum. If $f'(x) = 0$ and the second derivative is positive at this point, then the point is a relative minimum. If $f'(x) = 0$ and the second derivative is negative at this point, then the point is a relative maximum.

slope of the tangent line One way of interpreting the derivative of a function is to understand it as the slope of a line tangent to the function.

slope-intercept form The slope-intercept form for the equation of a line is $y = mx + b$, where m stands for the slope of the line and the line has its y-intercept (the place where the line crosses the y-axis) at the point $(0,b)$.

standard form The standard form (sometimes also called general form) for the equation of a line is $Ax + By + C = 0$, where A and B are not both zero.

substitution Integration by substitution is one of the most common techniques of integration, used to reduce complicated integrals into one of the basic integration forms.

tangent line The tangent line to a function is a straight line that just touches the function at a particular point and has the same slope as the function at that point.

trigonometric substitution A technique of integration where a substitution involving a trigonometric function is used to integrate a function involving a radical.

washer method A procedure for finding the volume of a solid of revolution by treating it as a collection of thin slices with cross sections shaped like washers.

Appendix

USING GRAPHING CALCULATORS IN CALCULUS

One important area that hasn't been addressed in the rest of this book is the use of modern technology. While it's possible to learn and understand calculus without the use of tools beyond paper and pencil, there are many ways that modern technology makes tasks easier or more accurate, and there are also ways that it can give insights that aren't as clear otherwise.

Of course, this appendix can't be exhaustive, but it will return to several of the examples from earlier in the book and show how you could apply graphing calculators to them.

Because the variety of different calculators available is tremendous, everything here will be done in general terms that should apply to any graphing calculator. For specific details about how to handle your own calculator, you should look at its manual, but this appendix can give you ideas about how that applies to calculus.

To keep things general and easy, this appendix usually just gives the calculator's decimal answers to four places, and anything you need to type into your calculator appears in **bold,** sticking as close as possible to the way things will appear on your calculator keyboard and screen.

Limits

Graphing calculators are ideal tools for evaluating limits. The more sophisticated models have this as a built-in function (consult your manual's index under "limits"), but on any calculator you can at least estimate most limits by looking closely at a graph of the function.

Example 2-3 Revisited: Evaluate $\lim_{x \to -3} \frac{x^2 - 9}{x + 3}$.

Graphing the function **y=(x^2–9)/(x+3)** on a calculator, you can visually estimate that for values of x near –3, the values of y on the graph are around –6. Most calculators won't even show the hole in the graph at this point without special effort on your part, since they plot individual points using decimal values that probably don't include exactly –3.

If you have trouble judging the *y* value visually, you can also use the zoom or trace functions on most graphing calculators to get a more accurate estimate. For instance, tracing this graph to an *x* value near –3, you find that when **x = –3.0159,** you have **y = –6.0159,** and from that it's not hard to guess that the limit is around –6.

Example 2-11 Revisited: Evaluate $\lim_{x \to 2^-} \frac{x+3}{x-2}$.

Most graphing calculators do a poor job of rendering graphs near vertical asymptotes, but if you know what you're looking for, you can easily get the information you need. In this case, when you graph **y=(x+3)/(x–2),** the screen should show the curve plunging downward as *x* approaches 2 from the left and veering upward as *x* approaches 2 from the right, possibly with a misleading vertical line where the calculator naively tries to connect the two parts. That downward spike as you near –2 from the left is your sign that the limit is –∞.

Example 2-14 Revisited: Evaluate $\lim_{x \to +\infty} \frac{x^3 - 2}{5x^4 - 3x^3 + 2x}$.

Graphing the function **y=(x^3–2)/(5x^4–3x^3+2x)**, you look out to the right-hand end of the screen to see what the height of the graph is for the larger values of *x*. Don't be fooled into thinking there's nothing there, it's just that the *y* value of the graph is so close to zero that it appears to overlap with the *x*-axis. If you trace the graph, you can find that when **x = 10**, you have **y = 0.0212**, so the limit seems to be 0.

Derivatives

The more sophisticated calculators available today can evaluate derivatives symbolically, giving the same exact values or functions that you can find by hand. Many calculators also have built-in features to numerically compute the value of the derivative of a function at a point. You can consult your calculator's manual for this. You can use any graphing calculator to get at least an approximate value for the derivative of a function at a point, and understanding how this works helps you understand what a derivative really is.

Example 3-17 Revisited: Find $f'(2)$ if $f(x) = \sqrt{5x^2 + 3x - 1}$.

Graph the function **y=√(5x^2+3x–1)** and use the trace feature to find the coordinates of a point just to the left of $x = 2$ (like **x = 1.9048, y = 4.7807**) and a point just to the right of $x = 2$ (like **x = 2.0635, y = 5.1459**).

Now use the traditional slope formula to find the slope of the line connecting these two points:

$$m = \frac{y_2 - y_1}{x_2 - x_1}$$

$$m = \frac{5.1459 - 4.7807}{2.0635 - 1.9048}$$

$$m = \frac{0.3652}{0.1587}$$

$$m = 2.3012$$

So from this, you can guess that the derivative is about 2.3, which is the same value found by hand in Chapter 3. You could also have been a little bit less careful, and quicker, by just keeping things to two decimal places and still gotten about the right answer.

The reason this works is because the derivative is just the limit that the slopes of the secant lines approach as the change in x goes to zero. By picking two x values with the change between them small, you found the slope of a secant line that's pretty close to the actual tangent line.

Another way to use a graphing calculator is to check answers you get by hand. The previous example could be seen that way, because when you worked it out by hand you got 23/10 and by calculator you got about 2.3. The more work done by hand the more likely most people are to slip. So, especially in a longer problem (like in the following example), verifying your work can be worthwhile.

Example 4-1 Revisited: Find the equation of the tangent line to the graph of $f(x) = \sqrt{x^2 + 3}$ at the point (–1,2).

When this problem was worked in Chapter 4, you found that $x + 2y = 3$ is the equation of the tangent line. Rearranging this to slope-intercept form, you can graph both **y_1=√(x^2+3)** and **y_2=–.5x+1.5** together, and the two graphs should appear to overlap near the point (–1,2). In fact, if you zoom in towards the point (–1,2), the closer you look, the more the two graphs should appear identical. The whole idea of a tangent line, after all, is that it should touch the function and have the same slope, so near the point of tangency it should be almost impossible to tell the two apart. If your graph hadn't turned out like this—if the tangent line hadn't touched the function at the right point, or if they didn't appear to have matching slopes there—you'd know something had gone wrong in your computations and could go back to check them over.

Example 4-6, Revisited: Find the maximum and minimum values of $f(x) = x^4 - 3x^3 - 1$ on $[-2,2]$.

Many graphing calculators have built-in features for finding maximum or minimum values of functions, but even without such a feature, graphing calculators make most extreme value problems easy. If you graph **y=x^4–3x^3–1** and make sure the viewing window includes x values from -2 to 2, you can see that the graph is highest at $x = -2$ (it also gets high on the right, but that's beyond your domain of $[-2,2]$). The lowest point seems to be near $x = 2$, but it's not immediately clear if it happens right at $x = 2$ itself. If you use the calculator's trace feature, you find out that the graph continues to decrease beyond $x = 2$, so the minimum value for your interval appears to happen at $x = 2$. You can now plug $x = -2$ and $x = 2$ back into the function to find that the maximum value is 39 and the minimum value is -9.

Example 4-9 Revisited: For $f(x) = \sin x + \cos x$ on $[0, 2\pi]$, determine all intervals where f is increasing or decreasing.

Graph **y=sin x + cos x** and make sure your viewing window includes x values from 0 to 2π. To make it easy on yourself, have the x-axis tick marks every $\pi/2$ units. (Most calculators have a feature that adjusts the viewing window to settings suitable for trig functions—also make sure your calculator is in radian mode rather than degrees.) It should be easy to see that the function increases for a short interval until the x value reaches $\pi/4$, then decreases until $5\pi/4$, and then increases the rest of the way to 2π, just as you found in Chapter 4—but this time with much less work!

Integrals

Some of the more sophisticated graphing calculators available today can evaluate both definite and indefinite integrals symbolically, quickly doing any of the problems you could work out by hand. However, most graphing calculators don't have this capability and therefore aren't much help with indefinite integrals. Most do have a built-in feature which numerically computes definite integrals, so check your manual for details.

One final case where graphing calculators and work done by hand can complement each other is finding areas bounded by curves, as in the following example.

Example 6-2 Revisited: Find the area of the region bounded by $y = x^3 + x^2 - 6x$ and the x-axis.

The actual integration involved in this problem is straightforward, but determining the limits of integration in the first place can be a nuisance. Graphing **y=x^3+x^2–6x** makes it clear that you need to integrate from $x = -3$ to $x = 0$, and then use the negative of the integral from $x = 0$ to $x = 2$ where the graph of the function lies below the x-axis.

Index

continued

E

F

G

H

I

L

M

N

O

Made in the USA
Monee, IL
04 November 2026

67377565R00085